R. K. Chhem · D. R. Brothwell

Paleoradiology

R. K. Chhem · D. R. Brothwell

Paleoradiology
Imaging Mummies and Fossils

With 390 Figures and 58 Tables

 Springer

Don R. Brothwell, PhD
Department of Archaeology
The University of York
The King's Manor
York Y01 7EP
UK

Rethy K. Chhem, MD, PhD, FRCPC
Department of Diagnostic Radiology and Nuclear Medicine
Schulich School of Medicine and Dentistry
University of Western Ontario
London Health Sciences Centre
339 Windermere Road
London, Ontario
N6A 5A5
Canada

Library of Congress Control Number: 2007936308

ISBN 978-3-540-48832-3 Springer Berlin Heidelberg New York

Springer is a part of Springer Science+Business Media
springer.com

Editor: Dr. Ute Heilmann, Heidelberg, Germany
Desk Editor: Meike Stoeck, Heidelberg, Germany
Production: LE-TEX Jelonek, Schmidt & Vöckler GbR, Leipzig, Germany
Reproduction and typesetting: Satz-Druck-Service (SDS), Leimen, Germany
Cover design: WMX Design, Heidelberg, Germany

Printed on acid-free paper   24/3180/YL   5 4 3 2 1 0

Foreword

The Radiologist's Perspective

It is my pleasure to write the foreword to this groundbreaking text in paleoradiology. Dr. Rethy Chhem is a distinguished musculoskeletal radiologist, and he is the founder of the Paleoradiologic Research Unit at the University of Western Ontario, Canada, and the Osteoarchaeology Research Group at the National University of Singapore. His special area of paleoradiologic expertise is the Khmer civilization of Cambodia, and his contributions to radiologic and anthropologic science have built bridges between these two not always communicative disciplines.

Dr. Don Brothwell is of course well known to the paleopathology community. He is something of an anthropologic and archaeologic polymath, having made important contributions to dental anthropology, the antiquity of human diet, and veterinary paleopathology, among others. His textbook, "Digging Up Bones" (Brothwell 1982), has introduced many generations of scholars to bioarchaeology, a discipline of which he is one of the founders. It is only fitting that this book is the work of a radiologist and an anthropologist, both of whom have experience in musculoskeletal imaging and paleopathology. For more than 100 years, diagnostic imaging has been used in the study of ancient disease. In fact, one of the first comprehensive textbooks of paleopathology, "Paleopathologic Diagnosis and Interpretation," was written as an undergraduate thesis by a nascent radiologist, Dr. Ted Steinbock (Steinbock 1976).

The advantages of diagnostic imaging in paleopathologic research should be intuitively obvious. Osseous and soft tissue may be noninvasively and nondestructively imaged, preserving original specimens for research and display in a museum setting. Not only will the original material, often Egyptian mummies, be preserved for future generations of researchers, but public enthusiasm will be fostered by the knowledge that we can see what is really underneath all those wrappings. Recent advances in computed multiplanar image display present novel ways to increase our understanding of the individuals, the processes of mummification and burial, and the cultural milieu in which these people lived. Unfortunately, although the potential of radiology has been recognized, the realization of collaborative effort has been inconsistent.

The earliest use of radiography in paleopathology was in the diagnosis of specific diseases in individuals, much as it is in clinical medicine today. Egyptian mummies were radiographed as early as 1896. Comprehensive studies of mummy collections were performed in the 1960s and 1970s, culminating in the exhaustive treatise by Harris and Wente, with important contributions by Walter Whitehouse, MD, in 1980 (Harris and Wente 1980). The usefulness of radiologic analysis of collections of such specimens led to the realization that diagnostic imaging has important implications in paleoepidemiology as well as in the diagnosis of individual cases.

Technical innovations in radiology have paralleled progress in paleopathology. We are now able to perform per three-dimensional virtual reproductions of the facial characteristics so that mummies do not have to be unwrapped, and we can now carry out "virtual autopsies" using three-dimensional computed tomography as a guide. We are now also using modern imaging technology to go beyond pictures. It is well established that radiologic and computed tomographic evaluation, in conjunction with physical anthropologic and orthopedic biomechanical data, may yield important biomechanical information in such studies as noninvasive measurement of the cross-sectional area of long bones to compare biomechanical characteristics in different populations such as hunter-gatherers and agriculturalists, and to study the mechanical properties of trabecular bone.

This textbook represents a significant advance in the effort to engage clinical physicians, especially radiologists and paleopathologists in a dialogue. Although there have been many such attempts in the past, they have for the most part dealt with specific imaging findings to diagnose disease in specific ancient remains. Chhem and Brothwell have given us the opportunity to go beyond this type of ad hoc consultation by presenting a systematic approach to the radiologic skeletal differential diagnosis of ancient human and animal remains. However, I believe that the intent of the authors is not so much to have paleopathologists interpret these finding in a vacuum, but rather to understand the capabilities of musculoskeletal radiologists, not only to assist with diagnosis, but also to offer information about the clinical setting in which these diseases occur and to suggest other appropriate imaging technology. For their part, musculoskeletal radiologists should be able to use this text to understand the context in which paleopathologists work, including taphonomic change, and to appreciate the rich legacy of diagnostic imaging in biological anthropology and archaeology.

Along with the authors, I hope that radiologists and biological anthropologists will use this textbook to translate both the radiologic and anthropologic idiom to better comprehend the other's potential for collaboration. Once we establish a common language, it will be easier to solve the diagnostic problems and dilemmas we share. Doctors Chhem and Brothwell are to be congratulated for taking that important first step.

Ethan M. Braunstein

References

Brothwell DB (1982) Digging up Bones (3rd edn). Cornell University Press, Ithaca

Steinbock RT (1976) Paleopathologic Diagnosis and Interpretation. Charles C. Thomas, Springfield, Illinois, p 423

Harris JE, Wente EF (1980) An X-Ray Atlas of the Royal Mummies. University of Chicago Press, Chicago, Illinois, p 403

Foreword

The Anthropologist's Perspective

The study of human paleopathology has benefited from the use of radiological methods for many decades. However, the use of radiological images and interpretative insights has in earlier years tended to be limited to medical professionals with expertise and experience in interpreting radiographic images as well as having access to the necessary equipment to produce radiographs in the hospitals where they worked. As the diagnostic value of radiology in the evaluation and diagnosis of disorders in archaeological human and nonhuman remains became more apparent, plain-film radiological facilities were established in many nonmedical centers where research on these remains was a central part of their scientific endeavors. With greater access to radiographic data on paleopathological specimens, biological anthropologists became increasingly competent in interpreting these images. However, there remain very important reasons why ongoing collaboration between radiologists and biological anthropologists in the analysis of paleopathological cases continues to be a valuable contribution to science.

One of the troublesome limitations of plain-film radiology is that three-dimensional anatomical features are projected onto a single plain. The inevitable superimposition that occurs can obscure important details of a radiographic image, adding to the challenge of interpretation. With the advent and widespread use of computed tomography (CT) radiological methods as an important diagnostic tool in clinical radiology, these methods began to be applied to archaeological remains. Among other advantages, CT imaging virtually eliminates the problem of superimposition. However, access to CT technology by paleopathologists, unless they are also radiologists, is often inconvenient or beyond the limited budgets of many researchers. This limitation in the use of CT imaging is changing as more facilities with CT equipment are available, including some in nonmedical research institutions. The remarkable power of CT imaging has made this mode of radiological investigation an important tool for the paleopathologist.

During my collaborations with radiologists in my own research on human skeletal paleopathology during the past 40 years, several issues have been highlighted. One is the need for better specimen positioning in taking radiographs of archaeological human remains. In clinical radiology, great attention is paid to the orientation of the anatomical site to be imaged relative to the axis of the X-ray beam. Clinical radiographic technicians receive careful training in the placement of the patient to be radiographed. Positioning of paleopathological cases of disease is often a helter-skelter arrangement in which little attention is paid to the anatomical relationship between multiple bones or the anatomical position relative to a living person. The emphasis is often on getting as many bones as possible on the X-ray film to save expense. Such a procedure does not lend itself to taking full advantage of the vast knowledge and experience of radiologists in the diagnosis of skeletal disorders.

Another problem is that in the burial environment, soil constituents often penetrate archaeological human skeletal remains and can pose real challenges in diagnosis, particularly for those inexperienced in recognizing these infiltrates. Soil infiltrates are denser than bone and appear as sclerotic areas in radiographic images.

These areas can be confused with antemortem pathology. Postmortem degradation of bone also occurs in the burial environment from both the acidic conditions commonly encountered in soil and the action of organisms, including bacteria, fungi and insect larvae, and plant roots. These destructive processes can mimic osteolytic pathological processes. Very careful attention to the fine details of the margins of destructive defects in bone is necessary to resolve the question of ante- versus postmortem destruction.

In interpreting radiographs of skeletal remains curated in museums, there is the further complication of distinguishing between substances added during museum curation of archaeological remains and antemortem pathological processes. For example, the glue used to repair breaks in older museum accessions can be very dense and create an appearance of a sclerotic response or a bone tumor in a radiograph.

These examples highlight the importance of collaboration between the clinical radiologist with an interest in paleopathology and the biological anthropologist in any study of archaeological remains, including mummified tissues and skeletal remains. Each discipline brings a specialized knowledge of the subject that maximizes the quality of the interpretation of radiological images from archaeological remains, both human and nonhuman.

Although collaboration between radiologists and biological anthropologists is an obvious strategy, the increasing use of radiology in the study of archaeological biological tissues calls for an explicit statement regarding the use of this methodology in research. As indicated above, the radiology of archeological remains poses special problems, and these need to be identified and resolved to ensure that radiographic data on such remains is interpreted correctly. There is a very real need for an authoritative reference work that will provide the insight from both anthropology and radiology as this relates to the use of radiological methods in the study of ancient evidence of disease.

I am very pleased to learn about the collaborative effort between Dr. Rethy Chhem, a skeletal radiologist, and Dr. Don Brothwell, a biological anthropologist, to produce a book on the radiology of archaeological biological tissues. Both are distinguished international authorities in their respective disciplines. In addition, both bring a depth of experience in the study of paleopathology that ensures careful coverage of the subject and new insight into the technical, theoretical, and interpretative issues involved in the application of radiology to the evaluation and diagnosis of abnormalities encountered in the analysis of human and nonhuman archaeological remains. I am confident that this book will be a major milestone in the study of disease in human and nonhuman archeological as well as paleontological remains.

Donald J. Ortner

Preface

This book arose from chance meetings and discussion between the two of us, one a radiologist and anthropologist (RC), the other a bioarchaeologist and paleopathologist (DB). The former expressed his interest in developing a scientific field that combined radiology with anthropology, especially bioarchaeology and paleopathology. The latter agreed completely that the subjects of radiographic techniques and the application of all aspects of medical imaging to the study of anthropological materials were sadly neglected. At the same time, both recognized that a publication was needed to show more clearly the considerable potential of paleoradiology. At this point, one of us (DB) expressed some uncertainty about finding the time (if not the mental strength) to contribute to the formation of this field. However, the extreme enthusiasm and persuasiveness of his friend and colleague (RC) resulted not in his withdrawal, but in discussing a joint plan of action. Such is the power of an enthusiastic colleague and a challenging project!

What follows in these pages is an attempt to introduce a new field of academic study that is concerned with the value of applying X-rays to a broad range of bio-anthropological materials, from human remains to other animals and even plants. We would emphasize that brought together in this way, it becomes a new field, even if components of the whole field have a much longer history. An entire chapter deals with the use of paleoradiology as a diagnosic method of ancient diseases. So in its entirety, the book is a pilot survey, an introduction to a broad-based subject that we feel is going to expand and interest a growing number of our colleagues, spanning human and veterinary radiology, anthropology (especially bioarchaeology), zoology, and botany. It is clear that at the present time, the literature relevant to this broad discipline is highly variable, and to some extent locked away in specialist publications. There is currently a strong bias toward human remains, both skeletal and mummified. We predict that this will change, and in particular we suggest that it will be employed increasingly in the field of zooarchaeology, where considerable numbers of bones and teeth are processed annually throughout the world and increasing attention is being paid to reconstructing the health status of earlier animal populations.

We sincerely hope that this introductory text on paleoradiology will stimulate interest in our colleagues, sufficient for them to ponder how they might contribute to this field in the future, or at least bring it to the notice of their colleagues or students. We do not see paleoradiology as a marginal and somewhat exotic occupation, rather one of considerable academic potential.

Rethy K. Chhem and Don R. Brothwell

Acknowledgements

We wish to express our great appreciation to our many friends and colleagues who have assisted in the preparation of this book in many ways. We sincerely hope that this list is complete, but if we have overlooked anyone by mistake, we ask for their forgiveness.

Supporting Rethy Chhem: Gord Allan, Ian Chan, Ghida Chouraiki, Eadie Deville, Jillian Flower, Jill Friis, Bertha Garcia, John Henry, Carol Herbert, David Holdsworth, Cheryl Joe, Stephen Karlik, Karen Kennedy, Jodie Koepke, Kyle Latinis, Luy Lida, Julian Loh, Liz Lorusso, Jay Maxwell, David McErlain, Wendy McKay, El Molto, Andrew Nelson, Jeremy O'Brien, Katie Peters, Christophe Pottier, Lisa Rader, Janine Riffel, Cesare Romagnoli, Frank Rühli, Roberta Shaw, Wang Shi-Chang, Vankatesh Sudhakar, Cynthia Von Overloop, Corie Wei, Jackie Williams, Deanna Wocks, Kit M. Wong, Eric Yap, Anabella Yim, and members of the Osteoarchaeology Research Group of Singapore.

Supporting Don Brothwell: Trevor Anderson, John Baker, Keith Dobney, Ben Gourley, Deborah Jaques, Simon McGrory, Theya Molleson, Naomi Mott, Sonia O'Connor, Terry O'Connor, Ian Panter, Jacqui Watson, Wyn Wheeler

We wish to thank GE Healthcare Canada for their support to the Paleoradiology Research Unit, and the Department of Radiology, London Health Sciences Centre and the University of Western Ontario, Canada.

Finally, but by no means least, we both wish to thank Sirika Chhem and Jade Orkin-Fenster, whose hard work and commitment in York during the summer of 2005 provided us with a wide range of digital radiographs for use in this book.

Contents

Chapter 3
The Taphonomic Process, Biological Variation, and X-ray Studies
Don R. Brothwell 55

Chapter 4
Diagnostic Paleoradiologyfor Paleopathologists
Rethy K. Chhem, George Saab,
and Don R. Brothwell 73

Chapter 5
Paleoradiology in the Service of Zoopaleopathology
Don R. Brothwell 119

Chapter 6
Normal Variations in Fossils and Recent Human Groups

Concluding Comments

List of Contributors

Don R. Brothwell, PhD (Editor)
Department of Archaeology
The University of York
The King's Manor
York Y01 7EP
UK

Rethy K. Chhem, MD, PhD, FRCPC (Editor)
Department of Diagnostic Radiology and Nuclear Medicine
Schulich School of Medicine and Dentistry
University of Western Ontario
London Health Sciences Centre
339 Windermere Road
London, Ontario
N6A 5A5
Canada

Richard N. Bohay, DMD, MSc, MRCD
Schulich School of Medicine and Dentistry
University of Western Ontario
London Health Sciences Centre
339 Windermere Road
London, Ontario
N6A 5A5
Canada

Ethan M. Braunstein, MD
Radiology Department
Mayo Clinic
Scottsdale
5777 East Mayo Boulevard
AZ 8505 Phoenix
Arizona, USA

Donald J. Ortner, PhD
Smithsonian Institution
Washington 20560
District of Columbia, USA

George Saab, MD, PhD
Department of Diagnostic Radiology and Nuclear Medicine
Schulich School of Medicine and Dentistry
University of Western Ontario
London Health Sciences Centre
339 Windermere Road
London, Ontario
N6A 5A5
Canada,

CHAPTER 1

Paleoradiology: History and New Developments

RETHY K. CHHEM

"... by far the greatest technical advance was made when radiology began to be used in the examination of anthropological and paleontological materials."... "The Roentgenological examination, moreover, has the great advantage in that it permits the investigator to examine bones without destroying them and to inspect mummies without unwrapping them."

(Sigerist 1951)

Paleoradiology is the study of bioarcheological materials using modern imaging methods, such as x-ray radiography, computed tomography (CT), magnetic resonance imaging (MRI), and micro-CT. The first x-ray study of human and animal mummies was performed by Koenig in 1896 (Koenig 1896), but to the best of my knowledge, the terms "paleoradiology" and "paleoradiologist" were coined much later by Notman, a radiologist at the Park Nicollet Medical Center in Minneapolis, in his article in the American Journal of Roentgenology in 1987 (Notman et al. 1987). Although "paleoradiology" etymologically means "ancient radiology," it is clear that when used within the context of paleopathology, the term defines without any confusion, the applications of x-ray tests to bioarcheological materials. Notman, in collaboration with a pathologist and an anthropologist from the University of Alberta, Canada, used radiographic investigation to study paleopathological lesions in two frozen sailors from the Franklin expedition (1845–1848) who perished in the Canadian arctic. As part of the investigation, Notman and his colleagues correlated the x-ray images of the two sailors with the results found at autopsy. Early publications on x-ray studies of mummies and skeletal remains include descriptive techniques, anatomy, and some of the paleopathology results.

The development of paleoradiology has, to a large extent, been dependent on the parallel development of radiology and medical imaging technology. Unfortunately, for the last 100 years, the lack of input from radiologists, particularly those with expertise in skeletal pathology, has hampered the development of a sound scientific foundation for diagnostic methods to assist paleopathology studies.

The availability of CT scanners in the early 1970s and the ongoing development of CT methods in the subsequent decades provided better visualization of the anatomy and of paleopathological lesions in mummies and in ancient skeletal remains. At the present time, the newer generations of CT scanners with their three-dimensional (3D) and surface rendering capabilities can create a 3D face reconstruction, or whole-body reconstructions of mummies. These have become extremely useful for anthropological studies museum displays and have attracted tremendous media attention. Despite these achievements, however, the role of CT in detecting ancient diseases is still not well established, largely due to a lack of clear diagnostic protocols. Most publications of CT deal with image acquisition of a whole body of a mummy, but without any tailored protocol designed to study a specific skeletal disease (O'Brien et al. 2007).

This textbook is an attempt to lay these crucial foundations for the development of a scientific method for diagnostic paleoradiology, providing paleopathology studies with the structure to develop as an evidence-based discipline. This is the approach of the Paleoradiology Research Unit at the Department of Diagnostic Radiology and Nuclear Medicine within the Schulich School of Medicine and Dentistry at the University of Western Ontario in London, Ontario (Canada). This endeavor is facilitated by the presence of renowned experts in medical imaging science at the Robarts Research Institute, a private medical research institute affiliated with the University of Western Ontario and located within the university campus.

Despite their laudable effort to acquire some radiological knowledge, many paleopathologists use what radiologists call "Aunt Minnie's" approach, which is to compare x-rays of a specimen with radiological images from textbooks to establish the final diagnosis. This approach has led to many errors in the interpretation of x-ray images simply because most x-ray patterns are not specific and a thorough differential diagnosis has not been discussed.

There is a widespread belief that radiologists who interpret x-rays of dry bone specimens are prone to mistakes because of the lack of understanding of taphonomic changes. This is, in my opinion, a logical fallacy, as once radiologists are made aware of these pitfalls, they will be integrated within the differential diagnosis of authentic paleopathological lesions. Who better to study disease than those who are intimately involved in its diagnosis? Therefore, the main reason for underdevelopment of paleoradiology is most likely that the paths of anthropologists and radiologists rarely cross. Bringing experts from these two separate scientific fields would, without any doubt, allow the establishment of evidence-based paleopathology.

This kind of close collaboration between clinical radiologists, medical imaging scientists, anatomists, pathologists, and bioanthropologists has allowed an intense cross-fertilization of ideas and forms a very strong interdisciplinary approach for the development of scientific paleoradiology and paleopathology disciplines at the University of Western Ontario. The imaging facility at Robarts Research Institute, in particular, has some of the most advanced medical imaging technology available, including multislice CT, micro-CT, high-field MRI, and magnetic resonance (MR) spectroscopy, all of which are useful for enhancing paleoradiological studies.

This chapter reviews the history and development of paleoradiology from its pioneering years to the present, when advanced medical imaging technology is used to investigate biological materials from archeological settings. The most recent developments of methods in paleoradiology are also reviewed, using the anatomical-clinical model (Boyer et al. 2003; Chhem 2006; Chhem and Ruhli 2004; Chhem et al. 2004), together with a radiological–pathological correlation model (Chhem et al. 2006; Notman et al. 1987). Radiological–pathological methods, as used in the clinical setting, are essential for a rational and objective interpretation of radiological findings in paleopathology, while keeping in mind any pitfalls caused by taphonomic changes.

Finally, we describe the ongoing advanced imaging investigation being carried out by our Paleoradiology Research Unit on some unique materials from the Royal Ontario Museum (ROM). The ROM has allowed members of our team access to its rare and precious collection of Egyptian mummies, among them the famous 3200-year-old mummified brain of Nakht, for which the first historical CT scans were performed on September 27, 1976.

1.1
Paleoradiography

The first documented paleopathological studies were recorded more that two centuries ago (Esper 1774) (Fig. 1.1). Much later, soon after the discovery of x-

rays by Roentgen in November 1895, but well before the official establishment of radiology as a medical specialty, x-ray study was used for nonmedical purposes to evaluate mummies of both humans and other animals, as well as to image ancient skeletal remains and hominid fossils (Böni et al. 2004; Koenig 1896) (Fig. 1.2). These studies were carried out primarily in Europe and in the USA (Albers-Schoenberg 1904; Culin 1898; Dedekin 1896; Eder and Valenta 1896; Elliot Smith 1912; Gardiner 1904; Gocht 1911; Gorjanovic-Kramberger 1902; Holland 1937; Koenig 1896; Londe 1897; Petrie 1898; Salomon 1921) (Figs. 1.3–1.5). In those early stages of x-ray technical development, radiological studies were performed on mummies for several reasons. X-ray images of the contents and the wrapping often were taken to distinguish authentic mummies from fakes, to evaluate the bone age, to detect skeletal diseases, and to search for burial goods. The most common geographic origins of mummies were Egypt and Peru, which served as materials for the first monography of paleoradiology that was published in 1930 (Moodie 1930) (Fig. 1.6). Occasionally, an x-ray study was performed to evaluate bones and

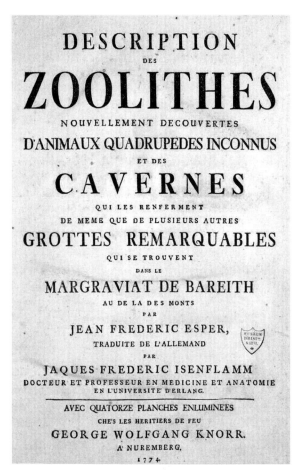

Fig. 1.1. Cover page of the first book on paleopathology

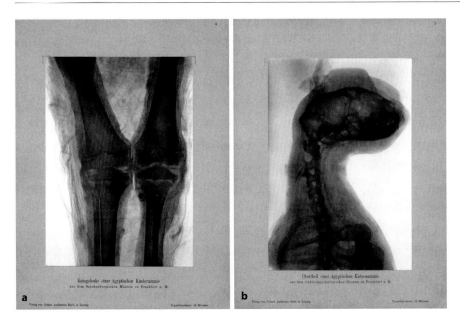

Fig. 1.2. a Koenig: Radiography of an Egyptian human mummy (1896). Reprinted with permission from Thieme, New York. **b** Koenig: Radiography of an Egyptian cat (1896). Reprinted with permission from Thieme New York

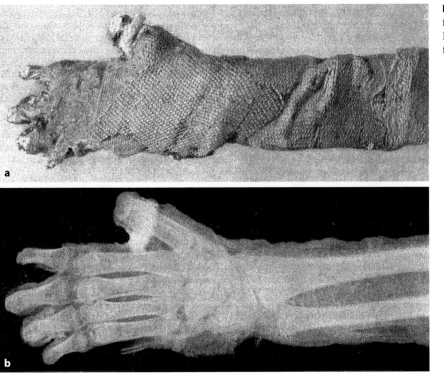

Fig. 1.3. a Londe: Mummy's forearm (1897). **b** Londe: Radiography of a mummy's forearm

teeth in paleolithic human fossils (Gorjanovic-Kramberger 1902). These early x-ray studies and results were published in French, German, or English in diverse scholarly journals (Böni 2004).

A historical review of the literature published within the first 25 years after the discovery of x-rays by Roentgen showed that paleoradiological studies were conducted by scientists from diverse backgrounds, including physicians and physicists, simply because there were no "radiologists" trained yet at that stage of x-ray development. For more information on this subject, a good source is Böni and his colleagues, who published a general review of the early history of paleoradiology (Böni et al. 2004) (Table 1.1).

Fig 1.4. **a** Londe: Fake mummy (1897). **b** Londe: Fake mummy

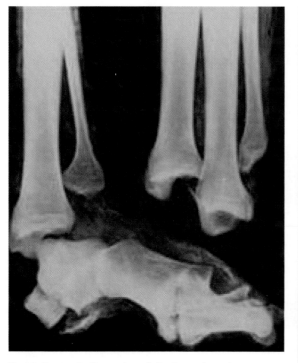

Fig. 1.5 Petrie: Radiography of the lower leg of a mummy (1898)

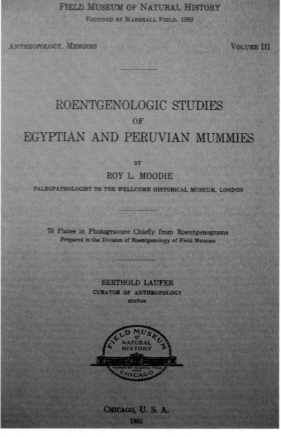

Fig. 1.6 Cover first book on x-ray study of mummies. Reprinted with permission from Field Museum Press, Fieldiana

Table 1.1. Early paleoradiology studies of mummies/skeletal remains/fossils

Author	Year	Study subject	Site
Koenig	1896	Human and cat mummies	Frankfurt, Germany
Holland	1896	Bird mummy	Liverpool, UK
Dedekind	1896/97	Egyptian mummies	Vienna, Austria
Londe	1897	Egyptian mummies Fake mummy	Paris, France
Leonard	1898	Peruvian mummies	Philadelphia, USA
Petrie	1898	Egyptian mummies	London, UK
Gorjanovic-Kramberger	1901	Hominid fossil	Vienna, Austria
Gardiner	1904	Egyptian mummies	London, UK
Albers-Schoenberg	1905	Egyptian mummies	Hamburg, Germany
Elliot Smith	1912	Egyptian mummies	Cairo, Egypt
Salomon	1921	Peruvian mummy	Berlin, Germany

1.2
Paleoradiology of Royal Egyptian Mummies

Because there is a widespread public and academic fascination with Egyptology, this section gives a detailed review of the history of paleoradiology of the royal Egyptian mummies, which has helped to shed light on the lives of ancient Egyptian rulers. The following section is a review of the literature related to the paleoradiology of royal Egyptian mummies (Chhem 2007) (Table 1.2).

1.2.1
1912 Thoutmosis IV

The first x-ray study of a royal Egyptian mummy was performed on Thoutmosis IV in 1903 by Dr. Khayat, an Egyptian radiologist. Thoutmosis IV was the 8th Pharaoh of the 18th Dynasty of Egypt, who ruled from 1400 to 1390 BC. The x-ray study provided the following information:

"...The left ilium (which was exposed in the embalming-incision) and the upper of the tibia (exposed in the broken right leg) was made, and other parts of the body were examined by means of the Roentgen-rays" (Elliot Smith 1912, p 44). "The epiphysis of the crest of the ilium was in process of union being united in the front but still free behind. This seemed to indicate that the body was that of a man of not more than 25 years.…..in Piersol's *Human Anatomy*, which was published three years after (in 1907) my report on

Table 1.2. Published x-ray studies on Egyptian royal mummies

Thoutmosis IV	1912 Elliot-Smith
Amenophis I	1933 Derry
Ramesses II	1976 Bucaille et al.[a]
	1979 Massare[a]
	1985 Bard et al.
	2004 Chhem et al.[a]
Tutankhamun	1971 Harrison[a]
	1972 Harrison and Abdalla[a]
	1976 Bucaille et al.[a]
	2003 Boyer et al.[a]
	2006 Shafik et al.[b]
Royal mummies	1972 Harris and Weeks
	1980 Harris and Wente
	1988 Braunstein et al.[a]

[a] Peer-reviewed journals
[b] Abstract

this mummy was written..." (Elliot Smith 1912, p 44). "In the skiagrams of this mummy, which were taken by Dr. Khayat in 1903, the epiphysis of the vertebral border of the scapula appears to be separate....but so far as it goes appearances support the low estimate of age, even if we accept Testut's date for the union of this epiphysis, ...and thereby extend the limit to 28 years. Judging from the texture of the bones as revealed by the x-rays, one would be inclined to admit that Thoutmosis IV might possibly have been even older than this." (Elliot Smith 1912, p 45).

1.2.2
1933 Amenophis I

Amenophis I (also known as Amenhotep I) was the 2nd Pharaoh of the 18th Dynasty, who is generally thought to have ruled for 20 years between 1526 and 1506 BC. His mummy was found by Victor Loret in 1898 in the Deir el-Bahri cache in the mortuary temple of Queen Hatshepsut in the Valley of the Kings. An x-ray study was done at the Cairo Museum on Saturday January 30, Tuesday February 2nd, and Thursday February 4th, 1932, after removal of the mummy from the coffin and cartonnage. Dr. Douglas Derry used x-ray findings to assess the age of the mummy: "The body proved to be that of an adult man. It is not possible to assign an age, except to say that all epiphyses are completely united and he is therefore above 25 years of age. So far as the teeth could be seen, they would were not unduly worn, nor are there any signs indicating advanced age such as loss of teeth or rarefaction of any of the bones, so that this king may have been about 40–50 years of age" (Derry 1933, p 47).

In his further evaluation of the x-ray study, Derry reported the following findings. "The cranial cavity appears to contain a diffuse mass, but whether this is the remains of the brain and membranes or whether it represents linen packing introduced by way of the nose, cannot be definitely decided, as the photographs do not show the condition of the ethmoid" (Derry 1933, pp 46–47). "The body has suffered considerable damage probably at the hands of the thieves. The right arm is bent at the elbow and the forearm is lying across the abdomen. There is a small amulet on the middle of the right arm, and towards the lower end of the arm there are two or three beads...". "The body cavity both chest and abdomen probably contains linen package" (Derry 1933, p 48).

1.2.3
1965 Royal Mummy Collection

In the spring of 1965, a team from the University of Michigan, in collaboration with Alexandria University in Egypt, was invited to undertake a paleoradiological study of skulls of ancient Nubian populations who lived near the Nile River. The Michigan project focused mostly on craniofacial variation studies. The radiological equipment used included a portable x-ray cephalometer using ytterbium-169 isotope with a half-life of 32.5 days, which allowed the equipment to be totally independent of a power source.

Following this first Nubian project, the Egyptian Department of Antiquities invited the same team, led by Dr. James E. Harris, Chairman of the Department of Orthodontics at the University of Michigan, to conduct an x-ray survey of the royal mummy collection of the Egyptian Museum. The project started in December 1967 with a radiographic study limited to the royal mummies' skulls. At that time, x-ray images of mummies were taken while they were still lying within their glass cases, to prevent any possible damage. However, the glass was found to contain lead, which severely degraded the images. In 1968, permission was given to remove the glass cases so that the mummies could be x-rayed in their wooden coffins, which resulted in far fewer artifacts than those caused by the glass cases. At this time the ytterbium source was replaced by a conventional x-ray machine using 90 kV, and, in addition to skull studies, a whole-body radiographic evaluation of the complete collection of royal mummies from the middle kingdom to the Roman period was performed.

The standard x-ray protocol included lateral and frontal views of the skull, the thorax, the pelvis, and the lower limbs. The data obtained during those multiple expeditions to Egypt form the basis of the publication of the Atlas of Royal Mummies by Harris and Wente (1980). The Atlas focused primarily on craniofacial variations and dental malocclusion, understandably, as the analysis of the data was conducted by a team of academic dental surgeons. The main limitation of this study was the lack of a specific x-ray protocol designed to study specific skeletal regions, as whole-body radiography was obtained for the survey. In lieu of a thorough analysis of x-ray data, apart from the study of craniofacial variations, there was a limited radiological inventory made available to potential mummy scientists, which was described in the preface of the Atlas. The preface stated that the reader was provided with "copies of x-rays from which he may draw his own conclusions and interpretations" (Harris and Wente 1980). This approach, although laudable, did not offer appropriate x-ray data for a paleopathological study of any of the royal Egyptian mummies. In addition, these data were not validated in the peer-reviewed literature until 1988 when 12 royal mummies were selected for paleopathological studies using x-rays as methods for disease detection (Braunstein et al. 1988). However, in 1973 the data were collated in a scholarly textbook, which became generally popular. Interestingly, this book entitled "X-raying the Pharaohs" (Harris and Weeks 1973) shed light on the context in which the radiological study of royal mummies was conducted, as revealed in the following quotes.

"We arrived at the museum each morning at 9 o'clock and, after signing the guard's register, proceeded upstairs to Gallery 52 where the mummies were displayed. While some of the staff began the task of setting the x-ray unit on its tripod, adjusting the transformers, and loading the film cassettes, two of us would decide on the mummies to be x-rayed that day. After a museum official and a guard arrived to oversee the work, the museum riggers would take one of the huge display cases, slide it into a narrow passage in the crowded room, and remove the leaded glass lid. Inside, the pharaoh lay in a solid oak coffin, covered with linen……..".

"The x-rays were taken, usually six to eight of each mummy, and the films rushed to a make-shift darkroom in a nearby hotel for developing. If they were acceptable, the workmen carefully returned the mummy to its case, sealed it, and prepared to bring another pharaoh to the unit. It was slow work and the average was four mummies a day…….".

"During the third season, a complete set of head-to-toe x-rays of each of the royal mummies was obtained. Some revealed important material for physicians and Egyptologists – information on pathological conditions, artifacts, and unusual techniques of mummification……"

"The expedition was confronted by two problems that first season. There was only a short time left after the problems of moving equipment from Aswan, and only frontal and lateral x-rays of the pharaohs' heads were obtained. Head-to-toe x-rays would have been preferable, since many of the mummies had never been unwrapped and held the possibility of revealing pathological conditions and, perhaps, artifacts. More importantly, the museum staff was justifiably concerned about the mummies, which, after 3000 years, were in fragile condition, and they asked that the x-raying be done through the glass display cases and no attempt be made to move the mummies about. This added greatly to the exposure time necessary to obtain prints, and, after several tests, it was discovered that the glass used in the cases was leaded. The x-rays penetrated with 3- to 5-minute exposures, but the resulting prints were generally foggy and lacking detail…….."

"In spite of the rather poor quality of the prints, everything went smoothly. The members of the expedition did not disrupt the work of the museum, nothing was damaged, and even the foggy x-rays proved interesting. The director of the museum, Henry Riad, realizing the problems the leaded glass cases had caused, and pleased with the results, invited the expedition to return the following year, when, he promised, the cases could be opened and clearer x-rays made" (Harris and Weeks 1973).

1.2.4
1968 Tutankhamun

The almost intact tomb of this young Pharaoh of the 18th Dynasty was found by Howard Carter's team in 1922 in the Valley of the Kings. In 1968, a team led by R.G. Harrison was granted permission to x-ray the mummy. Because the permission to remove the mummy from the tomb to Luxor hospital was not granted, an old portable x-ray machine (manufactured in 1930) was used. Parameters for exposure were selected based on a trial-and-error approach, and a set of test films was developed in a bathroom of the Winter Palace Hotel in Luxor. Once the quality of test films become acceptable, the rest of the x-rays films were sent to Liverpool for development.

Radiography of the skull provided information about the teeth that allowed an estimation of the age of Tutankhamun at between 18 and 22 years, although x-rays of the limbs also allowed an estimation of bone age, which suggested the age at death was 18 years. Radiographs of the thorax showed that he had not died of tuberculosis. Radiography of the abdomen and pelvis demonstrated mummification materials, but no evidence of any diseases. Finally, the estimation of the height based on measurements of the x-ray of the limbs showed that Tutankhamun's stature was 5 ft 6 in.

Radiography of the skull also showed two bony fragments in the skull cavity. The first may come from the ethmoid, but in a publication in 1971, Harrison stated that: "This piece of bone is fused with the overlying skull and this could be consistent with a depressed fracture, which has healed. This could mean that Tutankhamun died of a brain hemorrhage caused by a blow to his skull from a blunt instrument" (Harrison 1971). Using these same set of x-rays of the skull, Boyer et al. have more recently dismissed this murder hypothesis, because a review of skull and cervical x-rays did not bring any convincing evidence to support proposed "theories of a traumatic or homicidal death" (Boyer et al. 2003). A recent CT study (Shafik et al. 2006) confirmed the radiological results by Boyer et al. that had already demonstrated post-mortem fracture of Tutankhamun's skull, overturning the homicide theory of the king. It took almost 30 years to correct an erroneous medical diagnosis caused by poor interpretation of simple skull x-rays (Boyer et al. 2003; Harrison 1971). This is yet another example of the importance of having x-rays of mummies read and interpreted by an expert in the field, who in this instance would be a trained pediatric radiologist (Boyer et al. 2003).

1.2.5
1976 Ramesses II

Ramesses II, the third Pharaoh of the 19th Dynasty, ruled from around 1279 to 1213 BC. His original tomb was in the Valley of the Kings, but his mummy was moved to Deir el-Bahri, where it was discovered in 1881. More recently, on the recommendation of Dr. Bucaille, a French surgeon, the mummy of Ramesses II was sent to France for scientific study and arrived in Paris in September 26, 1976 (1 day later, the world's first CT of a mummy was performed in Toronto, Canada) and was returned to Cairo on May 10, 1977. Before its trip to Paris, this mummy was x-rayed in early 1976 in Cairo by Bucaille and his collaborators. The results of that study were presented at the annual meeting of the "Société Française de Radiologie in Paris April 26, 1976 (Bucaille et al. 1976). The mummy of Ramesses II was again radiographed at the Musée de l'Homme in Paris and the results were published in a monograph almost a decade later in 1985 (Bard et al. 1985). A xeroradiographic study of the mummy of Ramesses II was also carried out on December 20, 1976 and the results published in Brussels in French (Massare 1979) (Fig. 1.7). Massare, the sole author of the paper, claimed that Ramesses II suffered from ankylosing spondylitis, but in 2004 this diagnosis of spinal inflammatory disease was refuted by Chhem and his colleagues, based on a study of unpublished and limited x-ray materials provided by Fauré, one of the three radiologists involved in x-raying and interpreting the films in 1976 at the Musée de l'Homme in Paris. Chhem's alternative diagnosis was a diffuse idiopathic skeletal hyperostosis (Chhem et al. 2004). CT of the spine and sacroiliac joints would represent the gold standard with which to validate the diagnosis of diffuse idiopathic skeletal hyperostosis as established by this limited set of x-ray studies. Beyond spi-

nal disease, the x-ray study shows other findings such as calcification of the intracranial carotid arteries, periodontal abscesses, and a rotator cuff arthropathy (Bard et al. 1985; Massare 1979).

1.3 Paleo-CT

CT also uses x-rays, but records many images from different angles that are stacked together to show cross-sections of body tissues and organs. CT can provide much more detailed information than x-ray films, giving images of soft tissues and blood vessels as well as bone. In August 1974, two cerebral hemispheres were retrieved from an autopsy performed at the Medical Science Building at the University of Toronto, by an international multidisciplinary team sponsored by the Academy of Medicine, the University of Toronto, the Royal Ontario Museum in Toronto, and the Paleopathology Association in Detroit, Michigan (Hart et al. 1977). The first CT scan of Egyptian mummy material was performed on September 27, 1976 at the Hospital of Sick Children in Toronto on the preserved and desiccated brain of Nakht, a 14-year-old weaver who died 3200 years ago in Egypt (Lewin and Harwood-Nash 1977a, b) (Fig. 1.8). A CT study of another Egyptian mummy (Djemaete-sankh) was also performed from head to the hips (Harwood-Nash 1979). During the same period, the mummy of the famous Egyptian pharaoh Ramesses II was sent for a scientific investigation at the Musée de L'Homme, Paris. X-rays and xeroradiography studies were conducted by a French team of three radiologists: Drs. M. Bard, C. Fauré, and C. Massare. However, no CT investigation was performed during this project, which was begun in 1976, but during the same period, a mummy's brain and another mummy

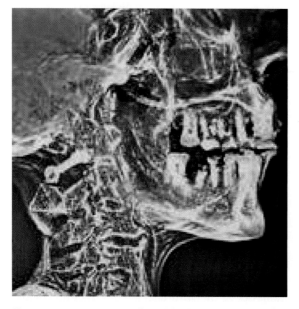

Fig. 1.7. Ramesses II. Xeroradiography, lateral skull and upper cervical spine (courtesy of Dr. Massare)

Fig. 1.8. First computed tomography (CT) of a mummified Egyptian brain (from ICRS Medical Science 1977)

were evaluated by CT in Toronto. The French team had missed a historical opportunity and the Toronto team became the first in the world to perform a CT scan of Egyptian mummies. The CT investigation included not only the study of a naturally mummified brain, but also whole-body imaging of the mummy. CT helps to assess not only the mummy's anatomy without the need of unwrapping, but also in the detection of amulets or paleopathological lesions. As the Toronto team used a first-generation CT scanner, the image resolution was poor. The overall morphology of the cerebral hemispheres and the ventricular outline were identified, the demarcation between the white and grey matters was faint, and a few post-mortem lacunae were identified. The thickness of each slice was 12 mm (Harwood-Nash 1979).

From this pioneering work to the current period, CT has been used to investigate many other mummies. Also, as CT technology has developed over time, the applications have expanded considerably (Table 1.3). More recently, micro-CT has been used to investigate mummy's materials (Fig. 1.10) and fossils (Chhem 2006; McErlain et al. 2004). Note that CT and micro-CT have been used on many bioarcheological materials other than mummies and skeletal remains. Although this subject is beyond the scope of this chapter, there are other sources for this information (Hohenstein 2004; McErlaine et al. 2004; Van Kaik and Delorme 2005).

1.4 Paleo-MRI

Magnetic resonance imaging (MRI) provides very detailed anatomical images of organs and tissues throughout the body without using x-rays, but MRI works by magnetizing the protons of the water mol-

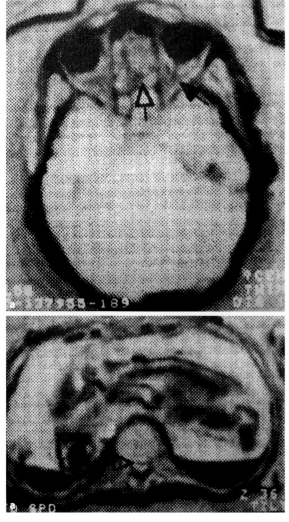

Fig. 1.9. a CT head. b CT abdomen. Both reprinted with permission from Lippincott, Williams, and Wilkins

Table 1.3. A 30-year-history of advanced medical imaging of mummies: milestones

Author	Year	Study subject
Lewin and Harwood-Nash	1977	CT mummy's brain (Nakht)-September 27, 1976
Harwood-Nash	1979	CT mummy's brain and whole-body
Marx and D'Auria	1988	3D Skull/Face
Magid et al.	1989	3D Entire skeleton
Nedden et al.	1994	CT Guided Stereolithography Head
Yardley and Rutka	1997	CT ENT (ear-nose)
Melcher et al.	1997	CT dentition-3D
Ciranni et al.	2002	CT skeleton/hand-tailored for arthritis
Hoffman et al.	2002	3D/Virtual fly-through "endoscopy"
Ruhli et al.	2002	CT-guided biopsy
Cesareni et al.	2003	3D-Virtual removal of wrapping
McErlain et al.	2007	Micro-CT of mummy's brain (Nakht)
Karlik et al.	2007	MR imaging and MR spectroscopy of mummy's brain (Nakht)

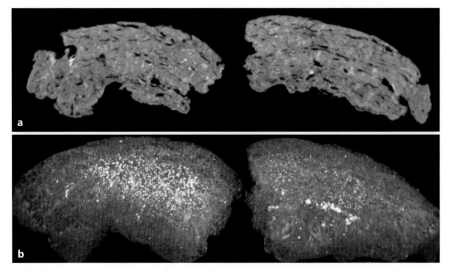

Fig. 1.10. **a** Micro-CT of Nakht's brain. **b** Micro-CT of Nakht's brain with maximum intensity projection

ecules within the body. As mummies are generally desiccated, MRI usually is not an efficient method to obtain the best images. Despite this, a few attempts were made in the early 1980s to investigate mummies using MRI. On July 23, 1983, an Egyptian mummy from the Minneapolis Institute of Arts was transported by a private jet to St Mary's Hospital, Rochester, Minnesota (USA) for MRI and CT studies. A body coil was used consisting of two saddle coils. Several sequences were carried out, including spin echo, inversion recovery, and free induction decay. Despite the diversity of pulse sequences used, it proved impossible to obtain an MRI signal or image (Notman et al. 1986). On June 21, 1991, another Egyptian mummy from the Egyptian Gallery of the Oriental Institute of the University of Chicago Hospitals was submitted for an MRI examination. Once again, no MRI images could be produced. MRI study has also been attempted in Minneapolis, Buffalo, and other medical centers with little success (Kircos and Teeter 1991). The first successful MRI study of mummies was conducted by Piepenbrink et al. at the University of Goettingen Institute of Anthropology in 1986. However, to obtain the image of a foot of a Peruvian child mummy, the authors rehydrated the samples with 20% aqueous solution of acetone for 18 days (Piepenbrink et al. 1986). This invasive method would not be approved by many museum curators, for whom the preservation of fragile and precious, mummified material is of utmost importance. The first successful MRI study of an ancient brain was actually performed in 1986 using a 0.15-T resistive magnet unit (Doran et al. 1986). The 8000-year-old brain was recovered from a swampy pond in Central Florida. MRI was possible because the brain was preserved in an aqueous environment. This MRI study allowed the identification of several anatomical structures, such as the occipital and frontal lobes, the cingulate gyrus, and the lateral ventricles. However, no MRI study of a desiccated mummy's brain had ever been conducted in the past until May 24, 2006, when the paleoradiological team of the Schulich School of Medicine and Dentistry at the University of Western Ontario obtained both MR images (Fig. 1.11) and spectra of a desiccated 3200-year-old brain from a 15-year-old Egyptian male (Karlik et al. 2007). This technological achievement may warrant the replacement of the term "paleoradiology" by "paleoimaging," although, given the underdevelopment of paleoradiology and the rarity and limitations of MRI study of bioarcheological materials, it might be wise to stick to "paleoradiology," at least for now.

1.5
Paleoradiology and Clinical Radiology: Historical Development

The use of x-rays dramatically expanded after their discovery by Roentgen in 1895. In 1897, Walsh published a book on the role of Roentgen rays in medical work, discussing the diagnostic possibility offered by x-rays in detecting diseases of bone including trauma, infection, tumor, congenital diseases, and rheumatoid arthritis. The first edition was followed by three others soon after. In the same period, a textbook focusing exclusively on "Disease in Bone and its Detection by X-rays" was written by W.H. Shenton in London, UK (Shelton 1911). This small didactic book described with great detail the radiographic findings in inflammatory bone diseases, tuberculosis, osteoarthritis, tumor, and osteomalacia. In the USA, Leon-

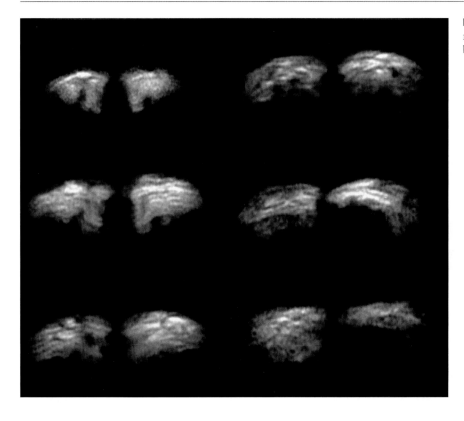

Fig. 1.11. Magnetic resonance imaging of Nakht's brain

ard of Philadelphia acknowledged that "the advent of the Roentgen method into the diagnosis of fractures has created the highest standard by which the results obtained in the treatment of fracture must be judged" (cited by Walsh 1907, p 186). Golding Bert urged "that radiography should be regarded as a subsidiary agent to diagnosis, and its evidence in cases of doubt and difficulty should be received with caution, and only after due interpretation by someone whose experience warrants his speaking with authority" (cited by Walsh 1907, p 198). Skiagraphy was coined in April 1896 by a 24-year-old medical student, Sydney D. Rowland, who served as editor of the world's first radiological journal called "Archives of Clinical Skiagraphy," which was devoted to new photography in medicine and surgery. In the preface of that first journal on radiology, Rowland wrote: "the object of this publication is to put on record in permanent form, some of the most striking applications of the new photography to the needs of medicine and surgery. The progress of this new Art has been so rapid that, although Prof. Roentgen's discovery is only a thing of yesterday, it has already taken its place among the approved and accepted aids to diagnosis...the first essays were of a rough and ready character; week after week, however, improvements have been made in the practical applications of the Art which I venture to call skiagraphy; and, at the present time, we are in the

position to obtain a visible image of every bone and joint in the body..." (Burrows 1986).

This brief review of the literature on radiology in the early phase of development of x-ray technology shows there was a considerable corpus of literature on radiology of bone pathology, indicating the availability of radiological expertise that paleopathologists could have used to enhance their attempts to detect skeletal lesions in mummies or dry bone specimens from archeological settings. The lack of actual interaction between radiologists and anthropologists/paleopathologists still plagues the methods and practice of paleopathology 100 years after the discovery of x-rays by Roentgen. One hundred and ten years after the first x-ray study of bioarcheological material performed by Koenig on a human and an animal Egyptian mummy, and despite the publication of paleoradiological articles in numerous and diverse scientific journals, there is still no single didactic paleoradiology book available to teach both the method and diagnostic approach of this discipline. Hence our endeavor to produce this book to fill this gap.

Although Harris et al. published two unique books on x-ray studies of royal Egyptian mummies, their purpose was not to teach paleoradiology, but instead to offer a kind of radiographic database of mummies for researchers in the field of Egyptology and mummy science (Harris and Weeks 1973; Harris and

Wente 1980). The Atlas of Egyptian and Peruvian Mummies, which was published by Moodie using the material from the Field Museum, Chicago, was the first book exclusively dedicated to x-rays of mummies (Moodie 1930). Again, the purpose of this atlas was to offer an inventory of human and animal mummies from the museum collection. The two classic textbooks on paleopathology (Aufderheide and Rodriguez-Martin 1998; Ortner 2003) contain a collection of ancient skeletal materials with pathologic lesions. However, x-ray images represent less than 5–10% of the total number of illustrations contained in these textbooks. Of all textbooks published on paleopathology, only one offers a didactic approach to diseases in the ancient skeleton using the "medical diagnosis" paradigm, which is that published by R.T. Steinbock in 1976 (Steinbock 1976).

Finally, after a review of the historical development of radiology for the last 100 years, we can define paleoradiology as a method that is used to investigate mummies and ancient skeletal remains. There is still the potential to expand the discipline to include the radiological exploration of other biological materials such as botanical fossils. According to Chhem (2006) paleoradiology can be divided into two main domains: anatomical paleoradiology, which covers morphological study and is useful for mummies and hominid fossils, and diagnostic paleoradiology. Borrowing from Ruffer's definition for paleopathology (Ruffer 1921), I would state that "diagnostic paleoradiology can be defined as the science of radiological detection of diseases, which can be demonstrated in human and animal remains of ancient time."

From this historical review, there is a need to make paleoradiology a more formal scientific discipline defined by sound methodology that should not be confused with techniques (Louiseau-Williams 2002), such as radiography, CT, micro -CT), or postprocessing of images such as 3D-surface-rendering. According to Louiseau-Williams, method is "the mode of argument that a researcher uses to reach explanatory statements." Lack of methodology, or at best weak methodology in paleoradiology has led to much confusion in this discipline for the last 100 years since its origin and early development. Considering this historical background, our book calls for increased collaboration between radiologists, paleopathologists and bioarcheologists, as well as other basic scientists in order to establish a solid evidence-based study of ancient human, other animal, and plant remains.

References

Albers-Schoenberg HE (1904) Röntgenbilder einer seltenen Koncherkrankung. Muench Med Wschr 51:365

Aufderheide AC, Rodriguez-Martin C (1998) The Cambridge Encyclopedia of Human Pathology. Cambridge University Press, Cambridge

Bard M, Fauré C, Massare C (1985) Etude radiologique. In: Balout L, Rouet C (eds) La Momie de Ramses II. Edition Recherche Sur les Civilizations, Paris, pp 68–75

Böni T, Ruhli FJ, Chhem RK (2004) History of paleoradiology: early published literature, 1896–1921. Can Assoc Radiol J 55:203–210

Boyer RS, Rodin EA Grey TC, Connoly RC (2003) The skull and cervical spine radiographs of Tutankhamen: a critical appraisal. Am J Neuroradiol 24:1142–1147

Braunstein EM, White SJ, Russell W, Harris JE (1988) Paleoradiologic evaluation of the Egyptian royal mummies. Skeletal Radiol 17:348–352

Bucaille M, Kassem K, Meligy RL, Manialawiy M, Ramsiys A, Fauré C (1976) Interêt actuel de l'étude radiologique des momies pharaoniques. Ann Radiol 19:5, 475–480

Burrows EH (1986) Pioneers and early years: a history of British radiology. Colophon, Alderney

Chhem RK (2006) Paleoradiology: imaging disease in mummies and ancient skeletons. Skeletal Radiol 35:803–804

Chhem RK (2007) Paleoradiological studies of Royal Egyptian mummies: history, development and future challenges. Abstract, VI World Congress on Mummy Studies, 20–24 February, Lanzarote

Chhem RK, Ruhli FJ (2004) Paleoradiology: current status and future challenges. Can Assoc Radiol J 55:198–199

Chhem RC, Schmit P, Fauré C (2004) Did Ramesses II have ankylosing spondylitis? A reappraisal. Can Assoc Radiol J 55:211–217

Chhem RK, Woo JKH, Pakkiri P, Stewart E, Romagnoli C, Garcia B (2006) CT imaging of wet specimens from a pathology museum: how to build a "virtual museum" for radiopathological correlation teaching. HOMO J Comp Hum Biol 57:201–208

Ciranni R, Garbini F, Neri E, Melai L, Giusti L, Fornaciari G (2002) The "Braids Lady" of Arezzo: a case of rheumatoid arthritis in a 16th century mummy. Clin Exp Rheumatol 20:745–752

Culin S (1898) An archaeological application of the Roentgen rays. Bulletin of the Free Museum of Science Department of Archaeology and Palaeontology, University of Pennsylvania 4:182–183

Dedekind A (1896) A novel use for the Roentgen rays. Br J Photogr 131

Derry DE (1933) An X-ray examination of the mummy of King Amenophis I. Trans ASAE 34:47–48

Doran GH, Dickel DN, Ballinger Jr WE, Agee OF, Laipis PJ, Hauswirth WW (1986) Anatomical, cellular and molecular analysis of a 8000-yr-old human brain tissue from the Windover arcaheological site. Nature 323:803–806

Eder JNL, Valenta E (1896) Versuche ueber Photographie mittlest der Roentgen'schen Strahlen von Regierungsrath Dr. J.M. Eder und E. Valenta. Herausgegeben mit Genehmigung des k. k. Ministeriums für Cultus und Unterricht von der k. k. Lehr- und Versuchsanstalt für Photographie und Reproductions-Verfahren, Wien. R. Luchner (W. Mueller), Wien

Elliot Smith G (1912) The Royal Mummies. Catalogue General des Antiquites Egyptiennes du Musée du Caire, 1912. L'Institut Français d'Archeologie Orientale, Le Caire. Reprinted in 2000, Duckworth, London

Esper JF (1774) Description des Zoolites Nouvellement Découvertes d'Animaux Quadrupedes Inconnus et des Cavernes qui les Renferment de Même que de Plusieurs Autres Grottes Remarquables qui se Trouvent dans le Margraviat de Bareith au Delà des Monts. (Translated from German by Isenflamm JF) Nuremberg, Knorr GW, 1774

Gardiner JH (1904) Radiographien von Mumien. The London Roentgen Society. 7. IV. Fortschr a. d. Gebiet der Roentgenstr 7:133

Gocht H (1911) Die Roentgen-Literatur. Zugleich An hang z Gochts Handbuch der Roentgen-Lehre. F. Enke, Stuttgart

Gorjanovic-Kramberger K (1901, 1902) Der palaeolithische Mensch und seine Zeitgenossen aus dem Diluvium von Krapina in Kroatien. Mittheilungen der Anthropologischen Gesellschaft in Wien 1901, 31:164–97 (4 plates); 1902, 32:189–216 (4 plates)

Harris JE, Weeks KR (1973) X-raying the Pharaohs. Charles Scribner, New York

Harris JE, Wente EF (1980) An X-ray Atlas of the Royal Mummies. University of Chicago University Press, Chicago

Harrison RG (1971) Post mortem on two pharaohs: was Tutankhamen's skull fractured? Buried History 4:114–129

Harrison RG, Abdalla AB (1972) The remains of Tutankhamun. Antiquity XLVI:8–18

Hart GD, Cockburn A, Millet NB, Scott JW (1977) Autopsy of an Egyptian Mummy, Can Med Assoc J 117:1–10

Harwood-Nash DC (1979) Computed tomography of ancient Egyptian mummies, J Comput Assist Tomogr 3:768–773

Hoffman H, Torres WE, Ernst RD (2002) Paleoradiology: advanced CT in the evaluation of nine Egyptian mummies. Radiographics 22:377–385

Hohenstein P (2004) X-ray imaging for palaeontology. Br J Radiol 77:420–425

Holland T (1937) X-rays in 1896. Liverpool Medico-Chir J 45:61

Karlik SJ, Bartha R, Kennedy K, Chhem RK (2007) MRI and multinuclear NMR spectroscopy of a 3200 year old Egyptian mummy brain. Am J Roentgenol (in press)

Kircos LT, Teeter E (1991) Studying the mummy of Petosiris: a preliminary report. News Notes Orient Inst 131:1–6

Koenig W (1896) 14 Photographien von Roentgen-Strahlen aufgenommen im Physikalischen Verein zu Frankfurt a. M. Leipzig: Johann Ambrosius Barth

Lewin PK, Harwood-Nash DC (1977a) X-ray computed axial tomography of an ancient Egyptian brain. ICRS Med Sci 5:78

Lewin PK, Harwood-Nash DC (1977b) Computerized axial tomography in medical archaeology. Paleopathol Newsl 17:8–9

Londe A (1897) Les rayons Roentgen et les momies. La Nature 25:103–105

Louiseau-Williams D (2002) The Mind in the Cave. Thames and Hudson, London

Magid D, Bryan BM, Drebin RA, Ney D, Fishman EK (1989) Three-dimensional imaging of an Egyptian mummy. Clin Imag 13:239–240

Marx M, D'Auria SH (1988) Three-dimensional CT reconstructions of an ancient human Egyptian mummy. AJR Am J Roentgenol 150:147–149

Massare C (1979) Anatomo-radiologie et vérité historique : a propos du bilan xéroradiographique de Ramsès II. Bruxelles-Medical 59:163–170

McErlain DD, Chhem RK, Bohay RN, Holdsworth DW (2004) Micro-computed tomography of a 500-year-old tooth: technical note. Can Assoc Radiol J 55:242–245

McErlain DD, Chhem RK, Granton P, Leung A, Nelson A, White C, Holdsworth D (2007) Micro-computed tomography imaging of an Egyptian mummy brain. Abstract, VI World Congress on Mummy Studies, 20–24 February, Lanzarote

Melcher AH, Holowka S, Pharoah M, Lewin PK (1997) Noninvasive computed tomography and three-dimensional reconstruction of the dentition of a 2,800-year-old Egyptian mummy exhibiting extensive dental disease. Am J Phys Anthropol 103:329–340

Moodie RL (1930) Roentgenologic studies of Egyptian and Peruvian mummies. In: Laufer B (ed) Anthropology Memoirs of the Field Museum, Vol III. Field Museum of Natural History, Chicago

Nedden DZ, Knapp R, Wicke K, Judmaier W, Murphy WA Jr, Seidler H, Platzer W (1994) Skull of a 5300-year-old mummy: reproduction and investigation with CT-guided stereolithography. Radiology 193:269–272

Notman DN, Anderson L, Beattie OB, Amy R (1987) Arctic paleoradiology: portable radiographic examination of two frozen sailors from the Franklin expedition (1845–1848). Am J Roentgenol 149:347–350

Notman NH, Tashjian J, Aufderheide AC, Cass OW, Shane OC III, Berquist TH, Gedgaudas E (1986) Modern imaging and endoscopic biopsy techniques in Egyptian mummies. Am J Roentgenol 146:93–96

O'Brien J, Battista J, Romagnoli C, Chhem RK (2007) CT imaging of human mummies: a critical review of the literature (1979–2005). Abstract, VI World Congress on Mummy Studies, 20–24 February, Lanzarote

Ortner DG (2003) Identification of Pathological Conditions in Human Skeletal Remains, 2nd edition. Academic Press, Amsterdam

Petrie WMF (1898) Deshasheh, 1897. Fifteen Memoirs of the Egypt Exploration Fund. The Offices of the Egypt Exploration Fund, London

Piepenbrink H, Frahm J, Haase A, Matthaei D (1986) Nuclear magnetic resonance imaging of mummified corpses. Am J Phys Anthropol 70:27–28

Ruffer MA (1921) Studies of Paleopathology of Egypt. University of Chicago Press, Chicago

Ruhli FJ, Hodler J, Böni T (2002) Technical note: CT-guided biopsy: a new diagnostic method for paleopathological research. Am J Phys Anthropol 117:272–275

Salomon F (1921) Roentgenbild eines peruanischen Mumientiels. Fortschr Roentgenstr 28:309–310

Shafik M, Selim A, Eisheik E, Abdel Fattah S, Amer H, Hawas Z (2006) The first multidetector CT study of royal mummy: King Tutankhamen. Abstract, Radiological Society of North America, Nov 26–Dec 1, 2006

Shenton WH (1911) Disease in Bone and its Detection by X-Rays. MacMillan, London

Sigerist HE (1951) A History of Medicine. Oxford University Press, New York

Steinbock RT (1976) Paleopathological Diagnosis and Interpretation: Bone Diseases in Ancient Populations. Charles C. Thomas, Springfield

Van Kaik G, Delorme S (2005) Computed tomography in various fields outside medicine. Eur Radiol 15:D74–81

Walsh D (1907) The Röntgen Rays in Medical Work, 4th edition. William Wood, New York

Yardley M, Rutka J (1997) Rescued from the sands of time: interesting otologic and rhinologic findings in two ancient Egyptian mummies from the Royal Ontario Museum. J Otolaryngol 26:379–383

CHAPTER 2

Paleoradiologic Techniques

2

George Saab, Rethy K. Chhem, and Richard N. Bohay

2.1
X-ray Imaging For Bioarcheology

X-ray imaging is used in three main types of human bioarcheological investigations. The first deals with the identification of anatomical structures that allow the determination of the stature, age at death, and gender. The second is to identify diseases in ancient skeletal remains and mummies. The last is the study of hominid fossils embedded in a burial matrix (Chhem and Ruhli 2004). In order to achieve these goals, bioarcheologists may need to undertake several steps. Bioarcheological materials can be submitted first to an x-ray investigation, and high-quality images can be obtained. The images are stored either on the traditional x-ray films or, more recently, on digital data supports. The ideal image analysis will be performed by radiologists with not only a qualification in musculoskeletal pathologies, but also equipped with an adequate and working knowledge of ancient bioarcheological materials. Alternatively, there can be collaboration between bioarcheologist and radiologist. These steps underline the need for adequate x-ray equipment and appropriate qualification in paleoradiology (Chhem 2006). X-ray studies have also been used to evaluate cultural material from archeological settings (Lang and Middleton 1997).

This chapter provides a general description of conventional and advanced imaging techniques suitable for bioarcheological applications. These include analogue and digital radiography, imaging physics, digital archiving, recent developments in computed tomography (CT), novel imaging methods, and three-dimensional specimen reconstruction techniques. A section on dental radiography has also been included. The imaging physics principles contained herein are not meant to be comprehensive, but rather to elucidate simple radiographic production factors that produce the best possible images. These factors exploit two important distinctions between bioarcheological and medical imaging applications: (1) the specimens do not move and (2) the total x-ray dose is less of a concern than it would be with a living subject. Emphasis is given to those technical factors that can be cont-rolled without costly upgrades. Care has been taken to avoid the use of physics and mathematical jargon in order to make this chapter accessible to readers without a background in radiology and physics.

X-ray equipment is available either in a hospital radiology department or in an anthropology department. In the former, one faces a few challenges, including the lack of specialized staff for taking radiographs of bioarcheological materials, but also the competition with clinical services. However, this is where one can have access to more advanced and costly imaging procedures such as CT scanning. Hospital x-ray equipments have also been used successfully to image 1-million-year-old slate fossils from the Devonian era (Hohenstein 2004). These plates of slate measure around 35 mm in thickness and contain a large variety of fossilized specimens including sponges, jellyfish, coral, mollusks, worms, and arthropods. The role of x-ray was to identify the fossilized animals, and to guide their exposure and preparation for pale-ontological study. Conversely, some x-ray equipment already available in an anthropology department may have a few limitations. Some types of x-ray equipment designed to study small specimens may not allow the study of an entire mummy or a large bone such as the femur or pelvis. In either department, mastering key concepts in x-ray imaging will help bioarcheologists to obtain the highest-quality image from their specimens. Beyond hospital facilities, a research imaging center offers the most cutting-edge technology (micro-CT scan and others) for the radiological assessment of bioarcheological materials. From this short review, bioarcheologists are facing technical, scientific, medical, and financial issues. Access to x-ray facilities, especially advanced imaging tests, represents the first challenge. Finding the expert to read and interpret the findings is also a major challenge. Diagnostic errors are common in paleopathology not only when x-rays are read by a radiologist with no specialized qualification in musculoskeletal pathologies, but also when the reader has no knowledge of the taphonomic processes that have altered the physical characteristics of skeletal specimens relative to those of the live clinical model. This stresses the value of a multidisciplinary approach to the radiological study of bioarcheological materials mentioned in the preface of this book.

2.2
Radiographic Production

2.2.1
Equipment Overview

Reduced to its elemental form, the x-ray imaging system consists of a high-voltage electrical supply, an x-ray tube containing a cathode and anode, and an image receptor (Fig. 2.1). The high-voltage power supply includes a series of transformers that amplify the electric inputs to meet the voltage requirements of the imaging system. This establishes a voltage across one end of the x-ray tube to the other. The peak kilovoltage, or kVp, can be manually set by the radiographer (Bushong 2004).

The x-ray tube is usually encased in an oil bath and lead housing. It is essentially a vacuum with two principle components: the cathode and the anode. The former consists of a coil of tungsten wire, much like

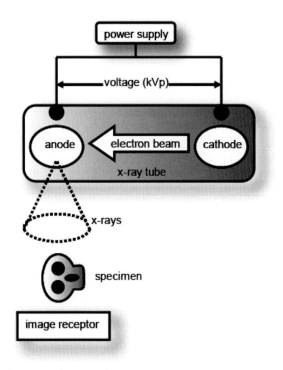

Fig. 2.1. A schematic of x-ray imaging system components. A power supply creates a voltage (peak kilovoltage, kVp) across the ends of an x-ray tube. This directs an electron beam from the cathode toward the anode. The electron beam produces x-rays that pass through the specimen to be captured at the image receptor, which is converted into a radiographic image

the filament of a household light bulb, surrounded by a focusing cup. Most x-ray tube cathodes actually have two separate filaments, each with an associated focusing cup. When the x-ray system is initiated, a current is passed through the tungsten filament. The high resistance of the filament causes it to heat with increasing current until it begins to boil off electrons from its constituent atoms. This process is called thermionic emission, and the product is a cloud of electrons. These electrons (or tube current) are accelerated across the tube by the kilovoltage and focused onto the anode by the focusing cup. Like the kVp, both the x-ray tube current in milliamps (mA) and the exposure time in seconds (s) are controlled by the radiographer.

The anode usually consists of a tungsten target mounted on a rotating surface, which is bombarded by electrons from the x-ray tube current. The rotating design allows a greater surface area to interact with the incoming electrons. This allows much higher tube currents to be used without damaging the target, as the heat is distributed along the entire target area. These interactions between electrons and the tungsten target produce x-rays. The area on the anode from which the x-rays are produced is the focal spot. X-ray tubes are usually equipped with two filaments and have two focal spots, one big and one small, for reasons that will become clear later in this chapter. X-rays are produced at the anode by two principle processes: Bremsstrahlung radiation and characteristic x-ray production (Johns and Cunningham 1984). The former is named for the German word for braking, producing x-rays with a continuous range of energies, with the maximum energy rays equal to the selected kVp. Characteristic x-rays occur at discrete energy levels. Characteristic x-rays result when an outer shell electron drops to fill the vacancy created when an inner shell electron is knocked from the atom. Both processes of x-ray production contribute to the x-ray emission spectrum (Fig. 2.2).

The anode is oriented to direct all of the produced x-rays towards the object or specimen to be imaged. A fraction of x-rays will emerge through the object, to be captured by an image receptor, which is any medium, analogue or digital, that converts incident x-rays into an image.

2.2.2
Portable X-ray Imaging Systems

There are currently several types of small, self-contained, shielded imaging systems available on the market that are well suited for bioarcheological applications. They are user-friendly and easy to maintain. These systems come with small focal spot sizes (20 μm

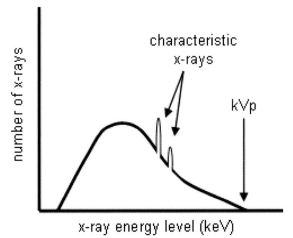

Fig. 2.2. The x-ray emission spectrum produced at the anode. The maximum energy of the x-rays in kiloelectron volts (keV) is equivalent to the peak kilovoltage (kVp) of the x-ray system. Characteristic x-rays occur at discrete energy levels, when electrons of the inner shells of atoms at the anode are ejected by the electron beam (the x-ray tube current)

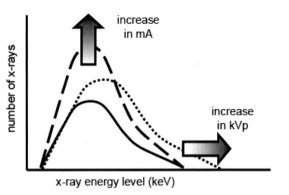

Fig. 2.3 Alterations in the x-ray emission spectrum. Characteristic x-rays have not been included in this figure, for simplicity. The shape of the spectrum (*solid line*) is affected by an increase in tube current (mA, *dashed line*), and an increase in kVp (*dotted line*)

is typical) and capacity for low peak kilovoltages, ranging from 10 to 40 kV, to optimize image quality. Optimization of these parameters is discussed in the subsequent sections of this chapter. These systems are designed with imaging areas in the range of 4.4–8.8 cm^2, appropriate for imaging small specimens.

2.2.3
X-ray Factors

For the purpose of this discussion, x-ray factors include the kVp, tube current (mA) and exposure time (s). These can be adjusted by the radiographer to modify the quantity and energies of the produced x-rays, thereby affecting the appearance of the final image (Bushong 2004). Increasing the tube current increases the amount of x-rays at each energy level, whereas increasing the kVp increases both the amount of x-rays at each energy level as well as the maximum x-ray energy (Fig. 2.3).

Higher-energy x-rays penetrate the object and register on the image receptor. Lower-energy x-rays do not contribute to the image and only add unnecessary radiation to the object, which is especially concerning for medical applications. Fortunately, the lower-energy x-rays can be selectively filtered in a process called beam hardening. The filtration occurs within the material of the x-ray tube itself and with additional layers of aluminum or copper placed in the path of the x-ray beam.

Why not maximize the kVp to produce high-energy x-rays, as shown in Fig. 2.3, and therefore produce a more penetrating beam? If the x-rays had sufficiently high energy, they would all penetrate the object regardless of its composition resulting in an image that would be completely homogenous and not very useful. This is discussed in more detail in section 2.3.1. The problem with high kVp also has to do with the way the x-rays interact with the object being imaged. X-rays interact with matter in many different ways, but only two are relevant for image production. These interactions are the photoelectric effect, which improves image quality, and Compton scattering, which reduces image quality.

2.2.3.1
The Photoelectric Effect

The photoelectric effect is a fundamentally important interaction between x-rays and matter (Fig. 2.4). It was first described in 1905 by Albert Einstein, who was recognized for this work with the 1921 Nobel Prize in physics. The photoelectric effect occurs when an incoming x-ray with energy equal to or slighter greater than that of a tightly bound inner shell electron of an atom, and an electron (referred to as a photoelectron) is ejected (Johns and Cunningham 1984). This leaves a vacancy in an inner electron shell that is quickly filled by an electron from an outer electron shell to stabilize the atom. The shift in energy levels from an outer to an inner shell causes excess energy to be emitted in the form of secondary x-rays, or in some instances the ejection of another electron called an Auger electron.

The photoelectric effect naturally requires incident x-rays to have at least as much energy as the binding energy of the electrons in the inner shell. The proba-

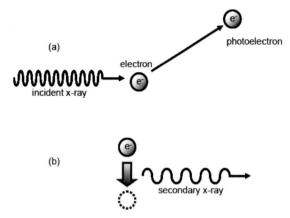

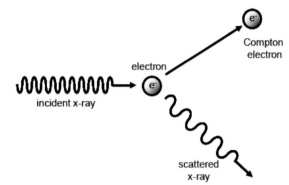

Fig. 2.4. The photoelectric effect. **a** An incident x-ray is absorbed by an inner shell electron (e^-) of an atom of the object being imaged. This causes an electron, called a photoelectron, to be ejected. **b** Secondary characteristic x-ray radiation is released after an electron from an outer electron shell drops down to fill the vacancy left by the photoelectron. The photoelectric effect produces the image contrast necessary for radiographic imaging

Fig. 2.5. The Compton effect. An incident x-ray interacts with an atom and ejects an electron from an outer shell. The electron is now called a Compton electron. The incident x-ray becomes a scattered x-ray, which continues at a different path with lower energy. The Compton effect reduces x-ray image quality

bility that the photoelectric effect will occur (which we want to maximize) is proportional to the atomic number of the absorbing material (which we cannot control) and inversely proportional to the x-ray energy (which we can control with kVp). This is a nonlinear relationship; modest decreases in kVp will produce large increases in the probability of this interaction, resulting in the image contrast necessary for image formation (Johns and Cunningham 1984).

2.2.3.2
The Compton Effect

The next interaction between x-rays and matter significant for x-ray imaging is the Compton effect, sometimes referred to as "Compton scattering". This phenomenon was first measured by Arthur Compton in 1922, earning him the 1927 Nobel Prize in physics. The Compton effect occurs between incident x-rays and lower-energy electrons, which reside in the outermost shells of an atom (Johns and Cunningham 1984) (Fig. 2.5). The incident x-ray collides with the electron, knocking it out of the atom, which is henceforth referred to as a Compton, or recoil, electron. The incident x-ray has transferred some of its energy to the Compton electron, but continues along a path anywhere from 0° to 180° from its original trajectory. Its energy decreases as a result and is now scattered, reducing the quality of the radiographic image.

The probability of the Compton effect occurring is proportional to the number of outer electron shells in the atom. The probability is also inversely proportional to the kVp, as was the case with the photoe-

lectric effect. The difference is that the relationship between the Compton effect and kVp is linear, thus small decreases in kVp produce small increases in x-ray scatter.

In summary, minimizing kVp increases the probability of the photoelectric effect, thereby producing differences in attenuation between anatomic structures in the object being imaged. Minimizing kVp also increases the probability of the Compton effect, which produces x-ray scattering to reduce image quality. Fortunately the increase in the photoelectric effect exceeds the increase in the Compton effect. Therefore, in considering both of these interactions, the radiographer is advised to select a kVp that is adequate for object penetration but kept at a minimal level.

2.2.4
Equipment Factors

In addition to x-ray factors, various components of the x-ray imaging system itself influence the appearance of radiographs. These include grids and radiographic film.

2.2.4.1
Grids

Scattered x-rays reduce image quality. The most common physical means of reducing scatter is a device called a grid. The grid is placed between the object and the image receptor to filter scattered x-rays so they cannot contribute to the image. Grids consist of alternating sections of material that x-rays cannot penetrate (radiopaque strips) and material through which x-rays can easily pass (radiolucent interspaced material strips). The strips are arranged to transmit only the x-rays directed toward the receptor that have not been scattered. One disadvantage of using a grid for medical applications is increased x-ray dose to the patient, as a higher tube currents and exposure times are required to make up for the x-rays lost to the grid. A disadvantage applicable to bioarcheological applications is the production of grid lines on the radiographic film, caused by the absorption of x-rays by the grid. This is a shortcoming that can be minimized by use of a reciprocating grid, which moves back and forth rapidly throughout the x-ray exposure, thereby decreasing grid lines (Bushong 2004).

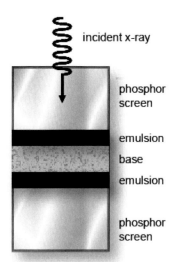

Fig. 2.6. Simplified cross-section of a screen-film image receptor system. The phosphor screens serves to absorb x-rays and emit visible light photons, which is recorded on the film emulsions. The emulsion-base-emulsion layers comprise the film

2.2.4.2
Radiographic Film

An image receptor is any medium that converts incident x-rays into an image. Film remains the most commonly used image receptor in radiography, although it is largely being replaced by computed and digital radiography (CR and DR, respectively) in the hospital environment (see sections 2.4.1 and 2.4.2, respectively). Most radiographic film consists of a base, which causes the film to be rigid, and an emulsion layer on both sides. The emulsion layer is a mixture of gelatin and silver halide crystals; this is the part of the film that creates the image. The main purpose of a dual emulsion film is to limit patient dose, a consideration less important for bioarcheological specimens. Most radiographic films are used in conjunction with an intensifying screen, a sheet of crystals of inorganic salts (phosphors) that emit fluorescent light when excited by x-rays. This serves to intensify the effect of x-rays during exposure of the radiographic film. Figure 2.6 is a schematic cross-section of a screen-film image receptor. For portability and durability, these are usually permanently mounted in cassettes (Bushong 2004).

When selecting a film-screen combination for bioarcheological radiography, it is important to consider the film speed. The faster the speed of the film, the thicker it will be, allowing for improved x-ray absorption and reduction in the necessary x-ray dose. However, this benefit is not as critical for bioarcheological specimens and it comes at the expense of resolution, therefore slow speed (thinner) film is optimal. Other important parameters to consider are single-emulsion films, to maximize resolution, and uniform small crystal size in the intensifying screen to provide high contrast and maximum resolution.

Proper handling and storage of radiographic film is very important. Film should be kept free of dirt, and bending and creasing films should be strictly avoided. The film is sensitive to light and radiation, so it must be stored and handled in the dark, away from sources of radiation, such as the x-ray imaging system. The storage area should also be dry and cool, preferably less than 20°C (68 F).

2.2.5
Geometry Factors

The geometric arrangement of x-ray equipment is an important determinant for image quality. Geometry factors include the size of the focal spot and its distance from the object and image receptor.

2.2.5.1
Focal Spot Size

Most general x-ray tubes are equipped with a small and a large focal spot. Recall from section 2.2.1 that the focal spot is the x-ray source, the area on the anode where the electron beam interacts to produce x-rays. A small focal spot provides greater image detail than its large counterpart because it casts the smallest penumbra, which is the area of blur at the edge of the image (Schueler 1998) (Fig. 2.7). One might wonder when a large focal spot would be ever required. It is used because the greater surface area for x-ray production minimizes heat production and the risk of the anode cracking or "pitting." Furthermore, a large focal spot enables the shortest possible exposure time to reduce blurring caused by patients who cannot remain still because of breathing problems or dyskinesia. This is not an issue for bioarcheological applications; a small focal spot should be selected with care not to pit or damage the anode by excessive use.

2.2.5.2
Source, Object, and Film Distances

As illustrated in Fig. 2.7a, even a small focal spot is not a precise point source. It has a finite area and will therefore cast an image penumbra, the area of blur at the edge of an image (Schueler 1998). The penumbra, along with image magnification, can be minimized by decreasing the distance from the object to the im-

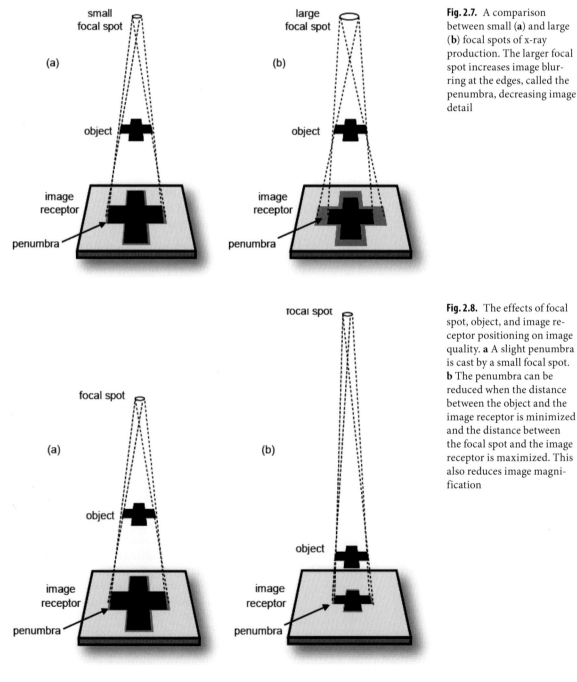

Fig. 2.7. A comparison between small (**a**) and large (**b**) focal spots of x-ray production. The larger focal spot increases image blurring at the edges, called the penumbra, decreasing image detail

Fig. 2.8. The effects of focal spot, object, and image receptor positioning on image quality. **a** A slight penumbra is cast by a small focal spot. **b** The penumbra can be reduced when the distance between the object and the image receptor is minimized and the distance between the focal spot and the image receptor is maximized. This also reduces image magnification

age receptor, and increasing the distance between the focal spot and the image receptor, as illustrated in Fig. 2.8.

2.2.6
Standard Radiographic Views

Bioarcheological specimens are often fragments, or isolated but intact bones, either cranial or postcranial. Postcranial bones refer to any bone other than those of the skull, which fall into four broad categories: long bones, short bones, flat bones, and irregular bones.

2.2.6.1
Cranial Bones

Despite the extensive use of CT in documenting skull pathology in clinical situations, conventional radiography of the skull is still widely performed today. Obtaining adequate radiographs can be difficult because of the complexity of skull anatomy, particularly when one is investigating fragmented bioarcheological specimens. Positioning a skull for standard radiographic views is often described in terms of the orbitomeatal line, sometimes denoted as the radiographic baseline. This refers to the conceptual line from the outer canthus of the eye to the center of the external auditory meatus (Ballinger 1982).The most common skull views are listed in Table 2.1 and illustrated in Figs. 2.9–2.14. These have been described previously in great detail in numerous texts and the interested reader is referred to the References section for a selection. A modified Caldwell is described rather than the original method because the modification is angled to further decrease the superimposition of the petrous ridges, which obscure the orbits. Anteroposterior views, as opposed to posteroanterior or modified Caldwell views, are selected when one wants to magnify the frontal structures, such as the orbits, which would be situated furthest from the image receptor. Recall that these magnified structures also have an increased penumbra and therefore increased blurring, as explained in section 2.2.5.2.

2.2.6.2
Postcranial Bones

Postcranial bones are classified as: long bones, including the femur, tibia, fibula, ulna, humerus, phalanges, and metacarpals; short bones, including the tarsals and carpals; flat bones, including the ribs, sternum, scapula, and skull bones; and irregular bones, including the vertebrae, and facial bones. These are imaged individually, in frontal and lateral projections (Ballinger 1982).

Table 2.1. Standard radiographic views of the skull. OML Orbitomeatal line, IOML infraorbitomeatal line, IR image receptor

Radiographic view	Skull OML position	Central x-ray orientation	Structures viewed best
Posteroanterior modified Caldwell	20° from vertical	Perpendicular to the IR	Orbits, frontal bone, anterior structures, frontal sinuses, nasal septum
Anteroposterior modified Caldwell	20° from vertical	Perpendicular to the IR	Magnified orbits, posterior view of the skull, frontal sinuses, nasal septum
Lateral	Interpupillary line perpendicular to the IR Midsagittal line parallel the IR	Perpendicular to the IR	Sella turcica, dorsum sellae, clivus
Waters	37°–40° to the IR	perpendicular to the IR	Maxillary sinus, orbits, zygomatic arches
Towne's	Perpendicular to the IR	Caudad 60° to the IR	Occipital bone, dorsum sellae, petrous ridges, foramen magnum
Basal	IOML parallel to the IR	perpendicular to the IR	Petrous bones, mandible, zygomatic arches, ethmoid sinuses, foramen magnum

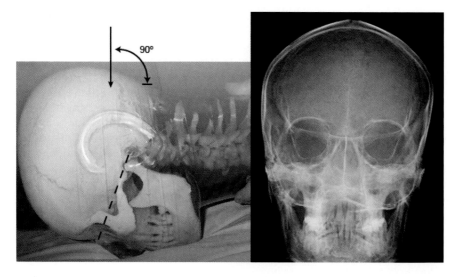

Fig. 2.9. Posteroanterior modified Caldwell view: positioning and resulting radiograph. The *straight arrow* signifies the direction of the central x-ray and the *dotted line* represents the orbitomeatal line, as described in Table 2.2. Images courtesy of Mr. John Henry

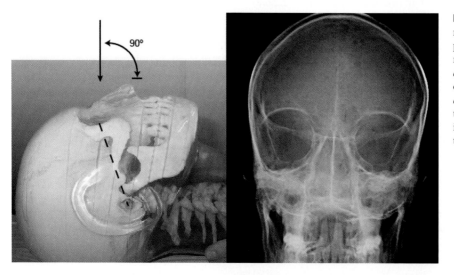

Fig. 2.10. Anteroposterior modified Caldwell view: positioning and resulting radiograph. The *straight arrow* signifies the direction of the central x-ray and the *dotted line* represents orbitomeatal line, as described in Table 2.2. Images courtesy of Mr. John Henry

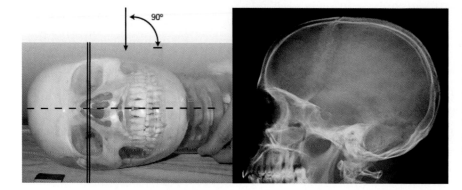

Fig. 2.11. Lateral skull view: positioning and resulting radiograph. The *straight arrow* signifies the direction of the central x-ray, the *double line* represents the interpupillary line, and the *dotted line* represents the midsagittal line, as described in Table 2.2. Images courtesy of Mr. John Henry

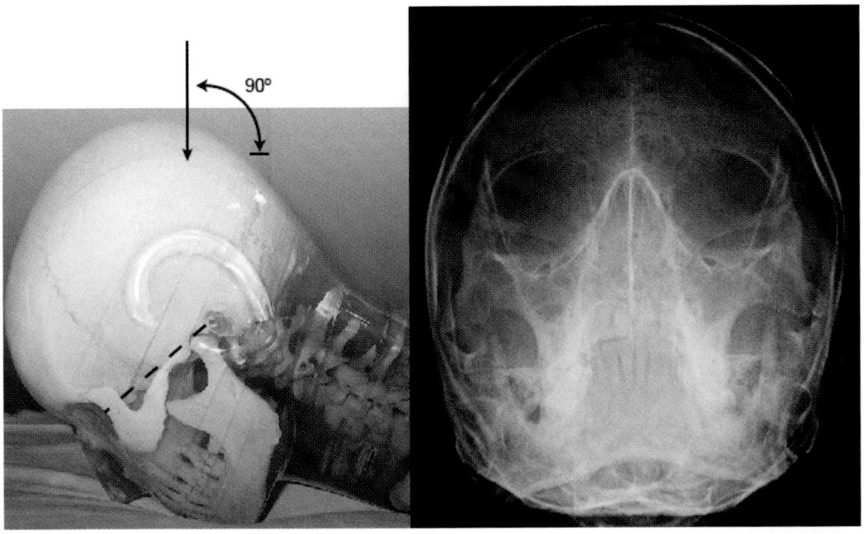

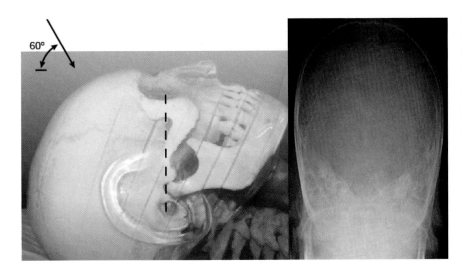

Fig. 2.12. Waters skull view: positioning and resulting radiograph. The *straight arrow* signifies the direction of the central x-ray and the *dotted line* represents the orbitomeatal line, as described in Table 2.2. Images courtesy of Mr. John Henry

Fig. 2.13. Towne's skull view: positioning and resulting radiograph. The *straight arrow* signifies the direction of the central x-ray and the *dotted line* represents the orbitomeatal line, as described in Table 2.2. Images courtesy of Mr. John Henry

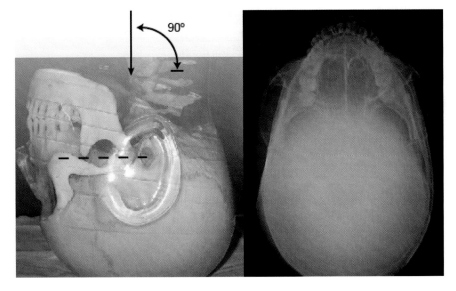

Fig. 2.14. Basal skull view: positioning and resulting radiograph. The *straight arrow* signifies the direction of the central x-ray and the *dotted line* represents the infraorbitomeatal line, as described in Table 2.2. Images courtesy of Mr. John Henry

2.2.7
Optimizing Radiographic Production Factors

The previous sections described how radiographers can optimize a wide variety of factors that influence appearance of the x-ray image. These included x-ray, equipment, and geometry factors; they are summarized for easy reference in Table. 2.2.

2.2.8
Quick Troubleshooting Guide

There will be times when x-ray systems fail. It is therefore a good idea to keep all of the service documentation in an easily accessible place. A troubleshooting flowchart is provided in Fig. 2.15 for quick reference. If all else fails, the best thing to do is to contact the manufacturer directly.

2.3
Image Quality

Radiographic images are said to be of high quality when they reproduce precisely the structures and

Table 2.2. X-ray imaging variables to optimize radiographs for bioarcheological applications

Radiographic production factors	Setting
X-ray factors:	
mAs (miliamperes × seconds)	High
kVp (peak kilovoltage)	High enough for object penetration, otherwise as low as possible
Equipment factors:	
Grid	Reciprocating
Screen–film combination	Slow speed, single emulsion, uniform crystal size in the intensifying screen
Geometry factors:	
Focal spot	Small
Source-to-image distance	Maximized
Object-to-image distance	Minimized

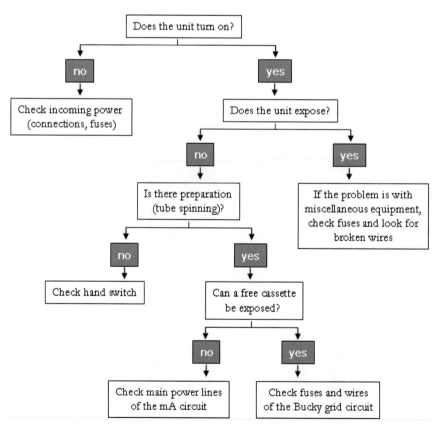

Fig. 2.15. Quick troubleshooting guide for equipment failure

composition of the object being imaged. There are many aspects to image quality, but here we discuss three of primary importance: contrast, resolution, and noise.

2.3.1
Contrast

Radiographic contrast is the difference in density between areas on an image. It is directly dependent on two separate factors: image receptor contrast and subject contrast. The former refers to the inherent properties of the film and processing factors. The latter pertains to the composition of the object to be imaged. X-rays do not penetrate an object equally because of differences in object density and atomic structure, resulting in images with good subject contrast. However, if the kVp were increased, more of the high-energy x-rays would be produced, decreasing the differences in the x-ray penetration throughout the object. Thus, the primary controller of subject contrast is kVp. This should be kept adequate for object penetration but set at a minimal level (Bushong 2004).

2.3.2
Resolution

Radiographic resolution can be described by many different and complex methods (Bushong 2004). Perhaps the simplest way is to consider resolution as the distinction between adjacent high-contrast structures on an image. Resolution is closely related to sharpness, which describes the abruptness between borders of adjacent high-contrast structures within an image. Resolution can be estimated by imaging an object of parallel line bars with alternating radiopaque and radiolucent lines. This object can be described in terms of line pairs per millimeter (lp/mm). For most diagnostic imaging examinations, a minimum resolution of 2.5 lp/mm is required, and preferably 5.0 lp/mm. The latter is in the range of standard films/screens used in hospital radiology departments.

Images with low resolution or sharpness are described as "blurry." One unavoidable cause of image blurriness is the structure of the object being imaged. Anatomical features that do not have well-defined edges, such as rounded objects, produce blurry edges on radiographs. Subject motion can also cause blurring, which could be problematic in bioarcheological imaging if the specimens roll or shift during the exposure. Another cause of blurriness is the selected film speed, described in section 2.2.4.2, and the penumbra effect described in section 2.2.5. The latter can be reduced by controlling geographic factors: increasing the distance between the x-ray source and the image, while minimizing the distance between the object and the image receptor (Fig. 2.8).

2.3.3
Noise

Noise is an undesirable characteristic of image quality that causes an image to appear textured or grainy. Noise can be caused by inherent properties of the image receptor. It could also originate from quantum mottle, a term given to the random interactions that x-rays have with the image receptor (Bushong 2004). The radiographer can reduce the noise caused by quantum mottle by increasing exposure (tube current multiplied by exposure time, mAs) to increase the number of x-rays that produce the image (Fig. 2.3), and decreasing the kVp so as to maintain the same density on the film. Noise is a greater problem with digital systems, as it is with any electrical system. Noise can be controlled with special software or by increasing the signal. The signal, determined by the kVp and mAs, is often quoted as a signal-to-noise ratio.

2.4
Advances in Radiography and Archiving

Radiographs have been recorded on film for over a century. Coupled with image intensifier screens and chemical processing after exposure, radiographic film has provided high-quality images that can be easily viewed on light boxes and archived. Despite the advent of digital imaging modalities such as CT and magnetic resonance imaging (MRI), radiography has not lost its importance or relevance, and has remained predominantly film-based. The reliance on film, however, is decreasing as DR methods challenge traditional screen-film image receptors. DR images equal or surpass the quality of film, and the high-speed electronic networks that are integral parts of healthcare and research infrastructure facilitate the transition and storage of digital media (Gallet and Titus 2005). The digital image receptors include CR, introduced in the early 1980s and, more recently, DR. Both CR and DR are briefly compared in Table 2.3. Although the image resolution of each is similar but less than that of film (5–8 lp/mm for screen-film at 400 ASA speed), the ability to "window" or dynamically alter the contrast of the image greatly enhances image quality and therefore facilitates interpretation. The specifications of the monitor used to display the images are very important, including brightness, amount of ambient light, size of display, number of gray scales, and monitor resolution.

Table 2.3. A comparison between computed radiography (CR) and digital radiography (DR)

IR type	Processing procedure	Processing time	Typical resolution
CR	Exposure of cassette, transfer to CR reader, transfer to network	30–45 s	3–5 lp/mm
DR	Exposure of digital panel, images immediately available	5–20 s	3–5 lp/mm

2.4.1
Computed Radiography

CR is a process of producing digital radiographs using a storage phosphor plate in the x-ray cassette rather than film. These storage phosphors store the energy from the x-rays. During processing, the phosphor is stimulated with a laser, causing it to emit light. The light is captured and converted into electrical signals. Instead of the chemical image development necessary for film-based radiography, a CR reader extracts the electrical information to produce a digital image (Gallet and Titus 2005).

CR is relatively easy to implement. It is fully compatible with existing x-ray equipment designed for film processing. The main disadvantage is that CR requires several steps for processing: the exposed cassette is brought to the CR reader, data is transferred to a computer, and then the cassette is erased. This takes approximately half the time of conventional radiography methods.

2.4.2
Digital Radiography

As CR was growing in popularity, a new step in digital imaging became available: DR. DR technology consists of flat-panel detectors with integrated image readout. Unlike CR, where an exposed cassette has to be physically brought to the CR reader, DR provides rapid access to digital images. Images can be viewed as they are being exposed, with a quality comparable to CR. DR excels for applications where speed and image quality are paramount, and its use is rapidly becoming more widespread (Gallet and Titus 2005).

2.4.3
Picture Archiving and Communication Systems

Digital images must be stored and archived. Picture Archiving and Communication Systems (PACS) is a software and computer server method for image storage and retrieval that has the potential to eliminate ra-

diographic film. Digital images from all imaging modalities, including radiography, CT, ultrasound, MRI, and nuclear medicine, are transferred through the computer network to a PACS server, which archives the images in a local drive (De Backer et al. 2004). A copy of the data is also stored on a separate archive. Whenever a stored imaging study is requested, all relevant prior imaging is also immediately available. In addition to being a robust archival system, there are numerous advantages for image viewing. The software has many features for image analysis, including image contrast windowing, measurement tools, and three-dimensional image reconstructions for tomographic data. With proper security access and software, images can be retrieved from virtually anywhere worldwide, greatly facilitating international collaboration and peer-to-peer information access. This is particularly beneficial to scientific disciplines such as bioarcheology, where multidisciplinary involvement is essential.

2.5
Computed Tomography

The main limitation of radiography is that all the structures of the object are superimposed onto a single image plane, where extraneous structures may obscure important findings. Godfrey Hounsfield and Allan Cormack independently described an imaging technique to overcome this limitation. They shared the 1979 Nobel Prize in Physiology and Medicine for the development of CT. Unlike conventional radiography, CT produces distinct images from multiple planes (or "cuts") through the object (Bushong 2004). Accordingly, the word "tomography" originates from the Greek word "tomos" meaning "to cut."

The original CT scanners of the early 1970s had a small bore (gantry aperture) designed solely for head scanning. In 1976, large-bore, whole-body scanners were introduced, and CT became widely available for medical applications in 1980. Typical bore sizes are in the range of 70 cm, although some manufacturers now offer CT scanners with bores up to 90 cm at an increased price. Large bores may ease the sensation

Table 2.4. Comparison of the various computed tomography (CT) design types

CT design	Principle	Source	Detector	Scan time per image
First generation	Rotate-translate	Pencil-beam	Single	5 min
Second generation	Rotate-translate	Fan-beam	Multiple, linear array	30 s
Third generation	Rotate	Fan-beam	Rotating curvilinear array	<1 s
Fourth generation	Rotate	Fan-beam	Stationary 360° curvilinear array	<1 s

of claustrophobia for clinical applications, but unfortunately are usually not sufficiently large to scan an adult-sized mummy casket or bioarcheological remains within a coffin.

2.5.1
Four Generations of CT Scanner Designs

CT scanners, regardless of their design, share the same basic elements. The differences between early designs pertain to the equipment configurations, including the type of source x-rays, number of x-ray detectors, and the relative motion between the source/detector system and the object to be scanned (Bushong 2004). The CT image is essentially derived by processing a large number of x-ray projections acquired through the object systematically from different angles. An x-ray projection is a depiction of the various x-ray attenuations obtained linearly from one end of the object

to the other (Fig. 2.16). Most CT designs can be classified in terms of four generations, as summarized in Table 2.4.

First-generation CT scanners (Fig. 2.17) are characterized by a single pencil-shaped x-ray source and a single detector. Both sweep across the object in unison to obtain an x-ray projection. This sweeping across the object is referred to as a "translation step." Successive views are obtained by rotating the source/detector pair and obtaining another projection at that particular angle. Hounsfield's original system required 180° translations, each separated by a rotation angle of 1°. This is the simplest of all the CT scanner designs.

Second-generation CT scanners (Fig. 2.18) use the same translation-rotation steps employed by the first-generation scanners. The difference is that the second-generation scanners incorporate fan-beam radiation and multiple detectors. Because the fan-beam has a wide angle of radiation, the x-rays measured by each detector are at a slightly different angle, so each translation step generates more data compared to that of a first-generation scanner. This means that larger

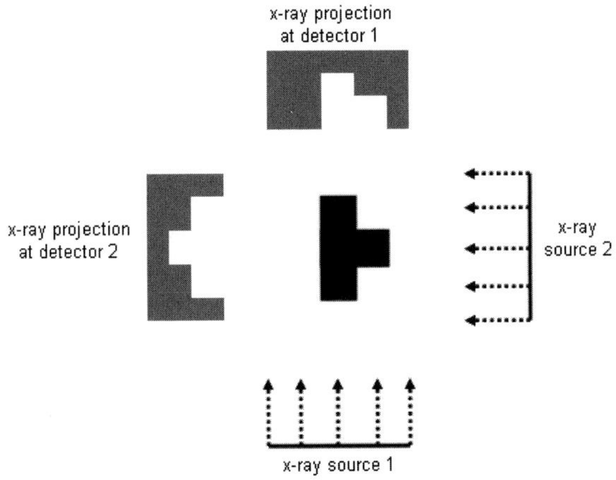

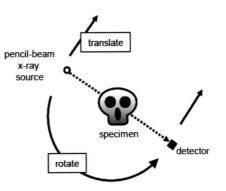

Fig. 2.16. X-ray projections. X-rays are transmitted from an x-ray source, penetrate the object, and are captured at the corresponding detector. A computed tomography (CT) image is created by computer processing of hundreds of x-ray projections

Fig. 2.17. A first-generation CT scanner. A physically coupled pencil-beam x-ray source and detector unit scan across the specimen (translation step) then rotate by one degree for another translation. This is repeated 180 times to acquire the raw data necessary for a CT image

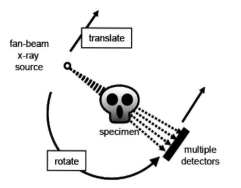

Fig. 2.18. A second-generation CT scanner. Like its predecessor, this design utilizes translation and rotation steps, but the fan-beam source and multiple detectors permit larger rotation increments and therefore faster scan times

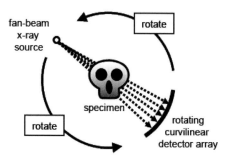

Fig. 2.19. A third-generation CT scanner. The translation step used in previous scanners is unnecessary in this design, as the source/detector pair rotates 360° around the specimem

rotation increments can be used, 5° or more, reducing the required number of translations, and consequently the scan time. The disadvantage to this design is that fan-beam radiation made it more difficult to correct for scatter, thus reducing image quality. This problem was addressed by the next generation of CT scanners.

Third-generation CT scanners (Fig. 2.19) also use a fan-beam source of radiation and multiple detectors, but in this design the detectors are arranged in a curvilinear array. This arrangement of detectors allows scatter to be reduced. There are more detectors and a greater fan-beam angle compared to the second-generation system, allowing radiation to encompass the entire cross-sectional area of the object. The translation step of image acquisition is no longer necessary. Instead, the source/detector pair simply rotates around the object. Once again, the scan time was greatly reduced since the previous generation. One potential disadvantage is the creation of ring artifacts, when each detector provides overlapping x-ray information, causing an apparent ring formation during the reconstruction. However, ring artifacts only occur when individual detectors are not properly calibrated, and are therefore easily corrected.

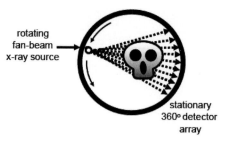

Fig. 2.20. A fourth-generation CT scanner. Multiple detectors completely encircle the specimen and only the fan-beam x-ray source rotates around the specimen

Fourth-generation CT scanners (Fig. 2.20) differ from the previous generation by having a stationary circular array of detectors completely encircling the object. Only the fan-beam x-ray source moves. This design is resistant to the ring artifacts that plague the third-generation scanners. One disadvantage is that the larger angle of source x-rays causes this design to be somewhat more sensitive to scatter than the previous generation. Both third- and fourth-generation CT scanners provide similar image quality and speed.

2.5.2
Spiral, Multislice, and Three-Dimensional CT

The first four generations of CT scanners acquire information for a single axial image at a time. The table (or couch) on which the object to be imaged is placed must be repositioned in a stepwise fashion to obtain data for subsequent images.

Spiral CT scanners have a fan-beam x-ray source and operate the same way as third-generation scanners. The distinction is that while the x-ray source is rotating in a circular fashion, the couch is simultaneously moving through the plane of the x-ray beam (Fig. 2.21). The amount the couch moves relative to the speed of x-ray source rotation is described by the term "pitch." Therefore the term "spiral" refers to the way the data is acquired, not the path of the x-ray tube. The spiral CT design allows for more rapid scanning, within a breath-hold for medical imaging applications (Bushong 2004).

If the number of detector arrays is doubled, two spiral sets of data (slices) can be acquired in the same amount of time previously required for a single slice. This is the configuration of a two-slice spiral CT scanner. It decreases the scan time by a factor of

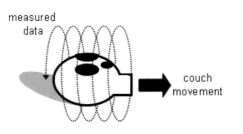

Fig. 2.21. Spiral CT scanning. The imaging data is obtained in the same fashion as a fourth-generation CT scanner, but the couch that the specimen rests on moves during the data acquisition, greatly improving the scan time

two, or allows twice the volume to be acquired in the same amount of time. Note that the term slice refers to the entire set of data, not a single transverse image, or "cut" through the object. Since the first multislice scanners were introduced in the early 1990s, developers have been working to push the limits of slice acquisition by continuing to add more detector arrays. Sixty-four-slice CT scanners are currently commercially available and this will surely continue to improve. Using advanced image reconstruction algorithms, the images can be reformatted into three-dimensional surface-rendered images that can be rotated.

2.6
Magnetic Resonance Imaging

In 1946, Felix Bloch of Stanford University and Edward Purcell from Harvard University, independently demonstrated that radiofrequency energy can excite hydrogen protons, elevating them to a higher-energy state. After a short time the protons return to their resting energy state, releasing electromagnetic energy in the process. This energy could be detected and recorded in what became known as nuclear magnetic resonance (NMR) spectroscopy (Nishimura 1996). Two decades later, Paul Lauterbur of New York State University produced the first magnetic resonance images of two test tubes (Lauterbur 1973). This was achieved by employing a magnetic field gradient to provide spatial information about the protons, thereby progressing from the single dimension of NMR spectroscopy to the two-dimensional MRI. In the late 1970s and early 1980s, several institutions and a rapidly growing list of manufactures began to produce images of in vivo human anatomy.

Today, MRI is nearly ubiquitous in hospital radiology departments, with numerous advantages over other imaging techniques including excellent soft-tissue resolution, variable contrast depending on the technique employed, and unlike CT, MRI does not

use ionizing radiation. Lauterbur was recently recognized for the discovery of MRI with the Nobel Prize for Physiology or Medicine in 2003.

2.7
Advanced Imaging Methods

Bioarcheological specimen imaging is by no means limited to radiography and conventional CT. New imaging techniques are being developed to provide a growing number of tools to evaluate the specimens.

2.7.1
Micro-Computed Tomography

One line of research focused on improving CT image resolution is the development of micro-CT (Holdsworth and Thorton 2002). These machines generally operate under the same physical principles as conventional CT scanners and are commercially available. The term "micro" indicates that the pixels on the resulting digital images are in the micrometer range. These systems have small imaging bores, as they were originally designed to study bone architecture and density in small animals. Depending on the bore size, these scanners are well suited for the study of individual bony specimens up to the size of an adult femur. The actual specifications can vary from system to system, but the smallest scanners will accept specimens from 1 to 4 cm in diameter, and can adjust its field of view and pixel spacing (by changing geometric magnification) to produce images with nominal pixel spacing between 11 and 50 µm. Other scanners are suited for specimens between 4 and 7 cm in diameter, producing pixel spacing between 25 and 100 µm. The largest micro-CT scanners accept specimens up to 14 cm diameter, with 100-µm pixel spacing. Any specimens bigger than this would have to be imaged in a regular clinical multislice CT. Micro-CT has been used to study ancient teeth and showed great anatomical details (McErlain et al. 2004).

2.7.2
Coherent-Scatter CT

For conventional imaging applications, scatter is an undesirable interaction between x-rays and matter because it degrades image quality. However, new imaging techniques are being developed based primarily on the scatter phenomenon (Batchelar and Cunningham 2002). Coherent-scatter CT (CSCT) produces cross-sectional images based on the low-angle (<10°) scatter properties of tissue. A diagnostic x-ray source and image intensifier are used to acquire scatter patterns under first-generation CT geometry. Following complex data processing, quantitative maps of bone-mineral content are obtained. This system will accept specimens up to 4 cm in diameter. CSCT may soon prove to be an important tool for densitometry of bioarcheological specimens.

2.7.3
Stereolithography and Fused Deposition Modeling

Advances are also being made in the way that the imaging data can be utilized. A set of CT imaging data for a bioarcheological specimen can be used to create a three-dimensional plastic representation. This is the principle behind stereolithography, also known as three-dimensional layering. A software algorithm divides the imaging data into thin layers and this information is output to a laser that acts on a tank filled with liquid photopolymer, a clear plastic that is sensitive to ultraviolet light. When the laser shines on the photopolymer, the plastic hardens. The laser thereby reproduces the specimen, layer by layer, until the model is complete.

Fused deposition modeling is an alternate method of three-dimensional printing that also recreates the specimen one layer at a time. Plastic material is supplied to a nozzle, which can move in horizontal and vertical directions under computer control. Layers are created as the nozzle is heated to melt the plastic, which hardens instantly after extrusion from the nozzle. Both stereolithography and fused deposition modeling allow access to the internal structures of bioarcheological specimens without damaging or even touching them (Recheis et al. 1999).

2.8
Dental Radiology

Methods of imaging the teeth and jaws of specimens will depend on whether the specimen is part of an intact skull, a fragment of tooth and bone, or simply a collection of loose teeth. In all instances, radiography will require an x-ray generator and, depending on whether the system is film or digitally based, a chemical processor or computer. This section will describe the basic technical factors in dental radiography, dental anatomy, and basic dental radiographic techniques of the teeth and jaws. An introduction to the radiographic appearance of dental caries, periodontal disease, periapical disease, and some abnormalities associated with impacted teeth are described at the end of this section.

2.8.1
Technical Factors

2.8.1.2
Dental Film

Dental film is available in various sizes and film speeds (Fig. 2.22). Unlike medical radiography, where most x-ray film is exposed with the use of intensifying screens, dental radiographs are exposed directly with x-rays. The purpose of the intensifying screen is to emit light when excited by x-ray photons. Because the screens are more sensitive to x-rays than the film, less radiation is required to create the image when a screen/film combination is used. However, this combination also results in decreased image resolution because the light emitted by the screen spreads as it travels toward the film, resulting in a larger area on the film being exposed than if x-rays were used alone (Fig. 2.23). Currently, dental x-ray film is available in D, E, and F speeds. Faster film is preferred for patient imaging because of the decreased radiation exposure. However, the slower ultra-speed, or D-speed, film has smaller silver halide crystals and is recommended when patient exposure is not a factor. Dental film is also available in single- or double-film packets. The use of double-film packets will provide two films with one exposure. The components of the x-ray film packet are arranged specifically and it is essential that the front (unmarked, white) surface of the film is placed adjacent to the material to be imaged. Within the packet, the film is surrounded by a light-protective black paper sleeve. Behind the film and paper sleeve is a thin, lead foil, which serves to decrease patient exposure and backscatter radiation, which would decrease the clarity of the image (Fig. 2.24).

2.8.1.2
Film Exposure

Dental film is usually exposed with a dental x-ray generator (Fig. 2.25), which has a fixed rectangular, or more often round, collimator, 6 cm in diameter. The role of the cone is to control the distance from the x-ray source to the skin in living patients. Dental units are usually fixed to the wall or ceiling of the room they occupy and regulations usually exist regarding placement of the generator and the type and size of barriers protecting personnel from x-ray exposure. Most modern dental x-ray generators have set kilovoltage (typically 60–70 kV) and milliamperage (typically 7–8 mA). Unlike medical x-ray equipment, the anode is stationary. Exposure time is variable, and depending on the unit ranges from a fraction of a second to a few seconds. Exposure time may be

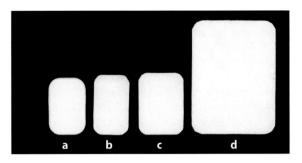

Fig. 2.22. Intra-oral dental x-ray film sizes: 0 (**a**), 1 (**b**), 2 (**c**), and 4 (**d**)

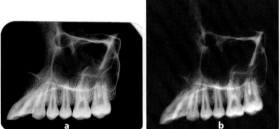

Fig. 2.23. a Image of the maxilla produced with size 4 (occlusal) dental film. **b** Image of the maxilla produced using a screen/film cassette. Image sharpness is improved using the nonscreen dental x-ray film (**a**)

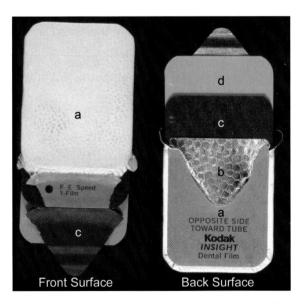

Fig. 2.24. Arrangement of the contents of the dental x-ray film packet. **a** Outer packet cover; **b** lead foil; **c** black paper wrap surrounding the x-ray film; **d** the x-ray film

measured in seconds or pulses (1 s=60 pulses). Older x-ray units often allowed variation in mA and kVp. In simple terms, exposure time and mA essentially

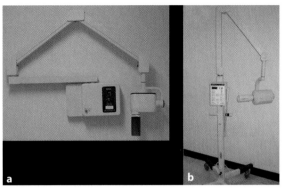

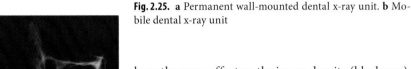

Fig. 2.25. a Permanent wall-mounted dental x-ray unit. **b** Mobile dental x-ray unit

have the same effect on the image density (blackness), where kVp primarily affects contrast. The ability to adjust kVp can be desirable when there is a need to change the contrast. High kVp techniques produce low-contrast images (many levels of gray) while low kVp technique produces high contrast images (few levels of gray between black and white). In imaging bone and teeth, high-contrast is often more desirable than low contrast, which is preferred in soft-tissue imaging. Modern dental x-ray generators have constant potential circuits that produce a more uniform, higher-energy x-ray beam throughout the exposure, resulting in decreased exposure times. These units describe the kilovoltage in terms of kV rather than kVp. Mobile dental x-ray generators are available and operate using a standard electrical supply. Mobile lead barriers are also available to protect personnel from occupational exposure. When barriers are not present, personnel should position themselves at least 6 feet (approx. 2 m) away from the x-ray generator and at an angle of between 90° and 135° to the x-ray beam.

2.8.1.3
Film Processing

Automatic dental film processors or manual processing will be required for dental film. Medical processors cannot handle the small film size and films will be lost within the processor unless attached to a larger "lead" film (Fig. 2.26). This method of processing is not ideal because the portion of the dental film attached to the lead will be unprocessed. Automatic dental film processors (Fig. 2.27) have the advantage of automatic control of time and temperature and produce dry films that can be handled immediately. Meticulous maintenance of processors is essential to good image quality. Film processing requires com-

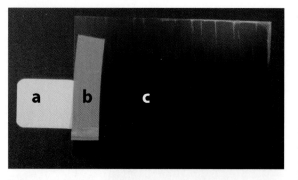

Fig. 2.26. Small dental film in a medical x-ray film processor can be accomplished by taping the dental x-ray film to a larger lead film with rubber tape. **a** Dental x-ray film. **b** Rubber tape. **c** Lead film

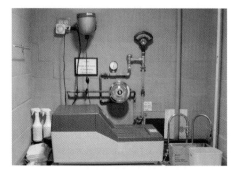

Fig. 2.27. Permanent, plumbed dental x-ray processor located in a dedicated dark room

mercially obtained chemical developer and fixer, and these must be changed and replenished regularly in order to obtain quality images. Depleted chemicals, and high or low temperature or time processing will negatively affect the quality of the image. In addition, most automatic processors employ a roller system to transport the radiographs through the processor. These rollers must be cleaned regularly in order to avoid films from becoming marked and soiled by contacting soiled rollers. Automatic processors must be connected to an electrical source and most must be plumbed. Where darkroom facilities are unavailable, a daylight loader attachment can be added to the processor, which permits film processing under daylight conditions. Manual processing is less costly and more mobile than an automatic processor. With meticulous technique, the images produced with manual processing will be of high quality. However, it is much more time consuming than automatic processing and films must be air dried in a dust free environment, which takes additional time after processing is complete. Manual processing is described in various dental radiography textbooks.

2.8.1.4
Digital Image Receptors

Digital image receptors are also available in dental imaging. The most common systems include photo-stimulable phosphor (PSP) plates (Fig. 2.28a), charge-couple devices (CCDs; Fig. 2.28b), and complimentary metal oxide semiconductor (CMOS) receptors. The diagnostic quality of the images is comparable to x-ray film. Advantages of digital imaging include postacquisition image manipulation, creating multiple high-quality copies, and eliminating the need for chemical processing. The disadvantages include cost, limited size of sensors, and nondicom proprietary software, which makes image sharing problematic in some applications. Also, CCD and CMOS sensors are thicker and inflexible, which can make sensor placement more difficult in intact jaws. For CCD and CMOS technology, the sensor is attached directly to a computer, or in wireless applications the computer must be in the vicinity of the sensor. The PSP plates are most similar to dental radiographic film in that they are thin and flexible. The exposed plates are taken to a central scanner that is attached to a computer. Digital images can also be created from dental film using a scanner with a transparency adapter.

2.8.1.5
Film Mounting and Storage

There are various methods of organizing and storing dental radiographs (Fig. 2.29). Commercially available film mounts are available in a variety of groupings based on typical dental patient examinations. Film mounts are opaque and allow for rapid removal and replacement of a film. Often, however, when different film sizes and number of films are used the ideal grouping may not be available. Radiographs can also be secured onto acetate film sheets with tape. This system has the advantage of allowing various size film and numbers to be organized in various presentations. It is important to secure both top and bottom edge of the film to the acetate to avoid film loss. Coin envelopes can also be used for storing radiographs that are not needed for examination. Of course digital images are stored electronically and it is always advisable to maintain remote backup.

2.8.2
Basic Anatomy of the Teeth and Jaws

The upper jaw is called the maxilla and the lower jaw is the mandible. The permanent dentition is composed of 32 teeth – 8 incisors, 4 cuspids, 8 bicuspids, and 12 molars. The primary dentition has 20 teeth – 8 inci-

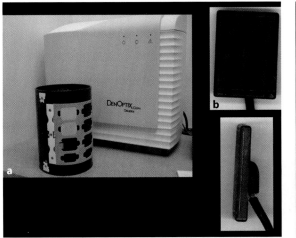

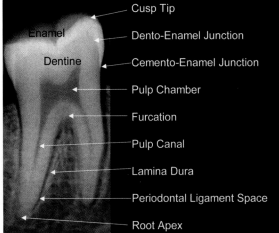

Fig. 2.28. a Photostimulable phosphor plate digital dental x-ray system. **b** Front and side view of a size-2 charge-couple dental sensor

Fig. 2.30. Radiographic anatomy of a permanent tooth

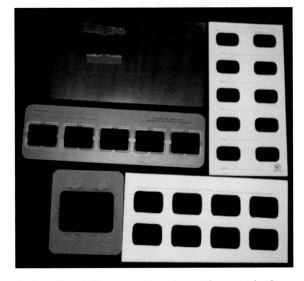

Fig. 2.29. Dental film mounting systems. The top right shows processed x-ray films taped to an acetate sheet. The various plastic and cardboard film mounts that are available for various sizes of dental film are also shown

sors, 4 cuspids, and 8 molars. The stage of development of the dentition can be used to estimate subject age. Each tooth has a crown, which is the functional portion of the tooth that is exposed to the oral environment, and the root(s), which are attached to the alveolar bone of the jaws through periodontal ligament fibers. Figure 2.30 shows the basic radiographic anatomy of a permanent tooth. The central portion of the tooth contains the dental pulp, which is composed primarily of loose connective tissue and vascular tissue. In dry specimens this space will be empty. The dentine, a substance that has about the same ra-

diographic density as bone, surrounds the pulp, and overlying the dentine is the enamel. The enamel terminates at the cemento-enamel junction, which is located just above the normal terminal position of the alveolar bone, called the alveolar crest. Each root of a tooth is composed of a central root canal extending from the pulp chamber and which is surrounded by dentine. The outer surface of the root is covered by a thin bone-like substance called cementum. The cementum is not evident radiographically except in cases of excessive deposition, called hypercementosis, a relatively common variation.

There are two common systems of tooth identification. In the American system, each tooth is numbered sequentially from 1 to 32, beginning at the maxillary right third molar. With the international system, the jaws are divided into four quadrants. The maxillary right is quadrant one, the maxillary left is quadrant two, the mandibular left is quadrant three, and the mandibular right is quadrant four. Within each quadrant, the teeth are numbered from one to eight beginning at the central incisor (Fig. 2.31). With the international system the primary quadrants are numbered five through eight and each tooth within the quadrant is numbered one through five, beginning with the central incisor. Primary teeth can also be identified by lower-case alphabet letters a–e. The surfaces of the teeth and jaws can also be identified (Fig. 2.32). The chewing surface of molars and bicuspids is called the occlusal surface, whereas the biting surface of anterior teeth is called the incisal surface. The surface adjacent to the cheek is called buccal, the surface adjacent to the tongue is lingual for lower teeth and palatal for upper teeth. The anterior surface of a tooth is mesial and the posterior surface is distal.

The incisor, cuspid, and bicuspid teeth usually have only one root, with the exception of the maxil-

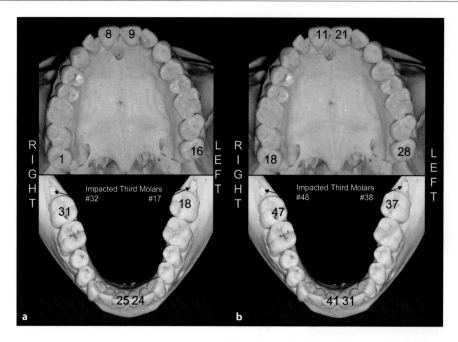

Fig. 2.31. Tooth numbering systems. **a** American; **b** International

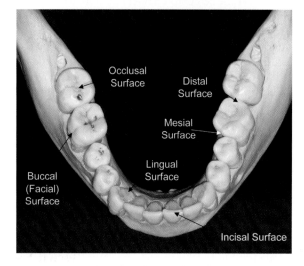

Fig. 2.32. Terminology describing surface of the teeth and jaws

lary first bicuspid, which usually has a buccal and palatal root. The maxillary molars typically have three roots – a large palatal root, and the smaller mesiobuccal and distobuccal roots. Most mandibular molars have a mesial root and a larger distal root.

Primary teeth can be described in terms similar to the permanent teeth. In general, primary teeth are smaller than their permanent successor and the enamel thickness is relatively thinner than in permanent teeth. This fact leads to a more rapid progression of dental caries in primary teeth as compared to permanent teeth and to the finding that primary teeth, near exfoliation, often show severe dental attrition.

The anatomy of the dentition is complex and beyond the scope of this book. Textbooks of dental anatomy are extremely valuable in the identification of teeth.

2.8.2.1
Basic Dental Radiography

Before describing some simple techniques for imaging the teeth and jaws of paleontological and archeological specimens, a basic description of the techniques used in routine dental radiography is described. There are three basic dental radiography techniques: the bitewing radiograph, the periapical radiograph, and the occlusal radiograph.

Bitewing Radiography

In bitewing radiography, the front surface of the x-ray film is placed adjacent and parallel to the lingual (palatal) surface of the maxillary and mandibular crowns of the teeth. The central ray of the x-ray beam is directed perpendicular to the radiograph. A bitewing tab is used to hold the x-ray film behind the teeth (Fig. 2.33). Two types of bitewing tabs are available; the most versatile is the stick-on type, which permits the film to be orientated in the usual horizontal fashion or vertically. The sleeve-type bitewing tab can only be used with a horizontal orientation of the film. When periodontal bone loss is present, a vertical bitewing is recommended, as it is more likely to capture the alveolar crest of the maxillary and mandibular teeth. In patients, the film is held in place by the patient, who occludes on a bitewing

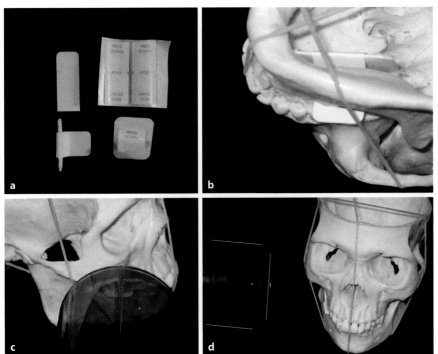

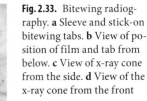

Fig. 2.33. Bitewing radiography. **a** Sleeve and stick-on bitewing tabs. **b** View of position of film and tab from below. **c** View of x-ray cone from the side. **d** View of the x-ray cone from the front

tab. The bitewing radiograph shows the crown of the maxillary and mandibular posterior teeth as well as a small portion of the alveolar bone nearest the cemento-enamel junction. The radiograph is usually made with size 2 film in adults and size 1 in children. The primary purpose of the radiograph is to examine the teeth for caries and to assess the periodontal status. A typical bitewing series is composed of an anterior and posterior bitewing radiograph made on each side. An ideal series will capture the distal of the canine and the distal surface of the most posterior tooth in the arch. The occlusal plane is centered horizontally and runs parallel to the top and bottom edge of the film. Ideally, there should be no overlap of the mesial and distal surfaces of adjacent teeth. In fact, due to the anatomy it is possible to create an "open" contact between the teeth radiographically, even when a tight contact exists anatomically. Fig. 2.34 shows examples of horizontal and vertical bitewing radiographs.

Periapical Radiography

Periapical radiography captures the entire crown, root, and surrounding bone (Fig. 2.35). They are generally made with size 1 or 2 film. While bitewings are usually only made for posterior teeth, where interproximal caries detection is dependent on radiographic methods, periapical radiographs are made of both posterior and anterior teeth. For adults, posterior teeth are imaged using a horizontally orientated size-2 film. Anterior teeth are imaged using a vertically placed size-1 film. Often the shape of the palate is too

narrow to allow the placement of the larger size-2 film in the anterior region without bending the edges of the film. When possible, it is ideal to maintain parallelism (parallel technique) between the film and the long axis of the teeth. In the maxilla and the anterior mandible in particular, the parallel technique requires that the film be situated further from the teeth. In some instances, the palate or floor of mouth does not allow parallel placement without losing the apices of the teeth on the image. In these circumstances, the bisecting angle technique should be employed. In this method the radiograph is purposely placed at an angle to the long axis of the teeth with the coronal edge of the film nearer the teeth and the apical edge of the film further from the teeth. The central ray of the x-ray beam is directed at an angle perpendicular to the bisector made by the angle between the long axis of the tooth and the x-ray film (Fig. 2.36). Distortion (unequal magnification of different parts of the structures) is more common with this method. Distortion causing a decrease in the vertical dimension of the teeth and jaws is termed foreshortening, while an increased length is termed elongation. Films are held in place with either plastic or Styrofoam film holders, which the patient bites on to hold the film in place (Fig. 2.37). A complete periapical series of the dentition typically includes: eight posterior periapical radiographs made with horizontally placed size-2 film (the premolar view captures the distal surface of the canine and the molar view captures the last molar in the quadrant), three vertically placed size-1 films

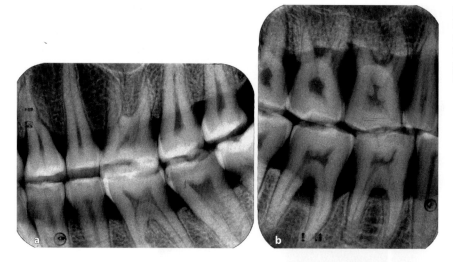

Fig. 2.34. **a** Horizontal bitewing radiograph made using a sleeve bitewing tab. **b** Vertical bitewing radiograph made using a stick-on bitewing tab

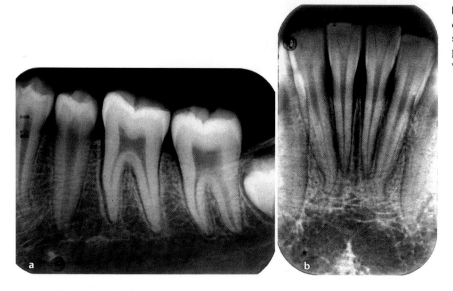

Fig. 2.35. **a** Posterior periapical radiograph made with a size-2 x-ray film. **b** Anterior periapical radiograph made with a size-1 x-ray film

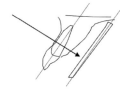

Parallel Periapical Technique

Bisecting Angle Periapical Technique

Fig. 2.36. Periapical radiography techniques

of the anterior mandibular teeth (centered over the left canine, incisors, and right canine), and five vertically placed size-1 films of the anterior maxillary teeth (centered over the left canine, left lateral incisor, central incisors, right lateral incisor, and right canine).

2.8.2.2
Occlusal Radiography

Occlusal radiography is a technique where the largest of the intra-oral film is placed parallel to the occlusal plane of the jaws. There are two types of occlusal radiographs – true occlusal views and standard occlusal views. The true occlusal view shows the teeth and jaws at a 90° angle to the standard periapical view. Standard occlusal views are basically periapical images of the teeth and jaws using a large film and a bisecting angle technique. For all of the occlusal views, the x-ray film is placed against the occlusal plane and the central ray of the x-ray beam is directed toward the center of the film. X-ray cone position can be confirmed by ensuring the lines marking the top and side of the x-ray cone bisect the front edge and the side

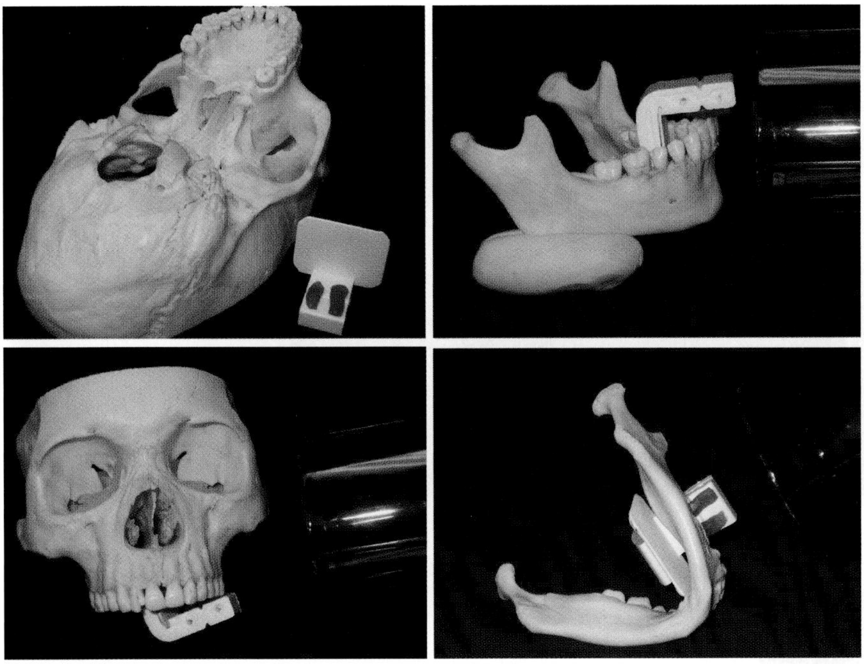

Fig. 2.37. Periapical radiography. Red rope wax on a Styrofoam holder to stabilize the film and holder in position. Modeling putty can be used to position and stabilize the jaw

edge of the x-ray film, respectively. A positive angulation of the central ray of the x-ray beam means that the x-ray cone is situated above the occlusal plane and angled downward toward the teeth. A negative angulation means the x-ray cone is situated below the occlusal plane and angled upward toward the teeth.

True Maxillary Occlusal (Vertex View)
The value of this radiograph in evaluation of the dental assessment is minimal in situations where all the teeth are erupted, but can be useful in assessing the position of unerupted and impacted teeth in the maxilla. It is also a useful view for assessing the buccal cortex of the maxilla (Fig. 2.38).

True Mandibular Occlusal
The true mandibular occlusal view (Fig. 2.39) will be useful in evaluating the buccal and lingual cortices of the mandible as well as the position of unerupted teeth and impacted teeth. In patients, this radiograph is often used to assess the soft tissue of the floor of the mouth where calcifications within the submandibular salivary gland duct can be observed radiographically. Usually, the film is placed so that the long axis of the film is bisected by the midsagittal plane. However, in some cases it is desirable to place the long axis of the film anteroposteriorly and image the left and right sides using separate films. This is particularly true when disease is present that has expanded the buccal cortex.

The next three views are best thought of as large format, periapical radiographs using a bisecting angle technique. These views are particularly useful when

the anatomy and bone surrounding the teeth is of interest and a larger coverage area is desired. These films are also useful in situations when the maxilla and mandible are articulated and opening is limited.

Anterior Maxillary Occlusal
This view (Fig. 2.40a) can be used to assess the location of impacted canines and posterior teeth as well as assess the maxillary buccal cortex. The x-ray beam is typically angled at +65° to the occlusal plane.

Lateral Maxillary Occlusal
This view (Fig. 2.40b) can be used to localize impacted teeth in the anterior maxilla as well as assess the alveolar process and maxillary sinus. The x-ray beam is angled +60° to the occlusal plane.

Anterior Mandibular Occlusal
This view (Fig. 2.41) is similar to the anterior mandibular periapical view, but provides an image of the entire tooth and anterior mandible. The x-ray beam is angled – 55° to the occlusal plane.

2.8.3
Specimen Imaging

2.8.3.1
Imaging Intact Jaws

Dental imaging will usually involve periapical and occlusal views. Periapical views will use size-1 film

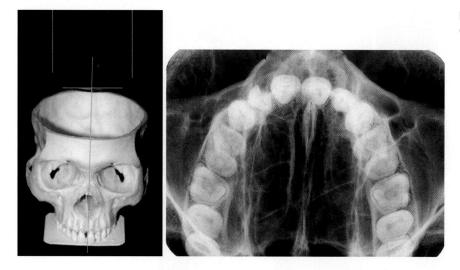

Fig. 2.38. Maxillary vertex occlusal view

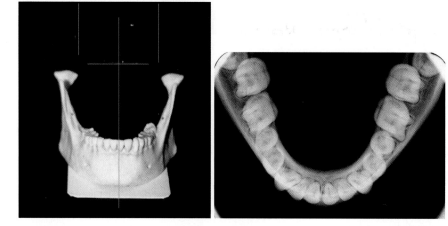

Fig. 2.39. True mandibular occlusal view

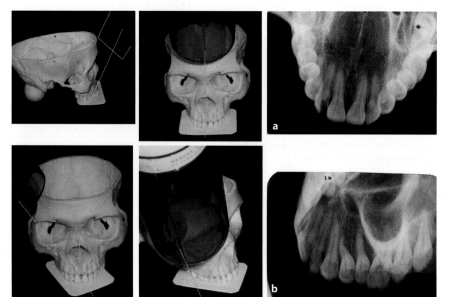

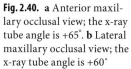

Fig. 2.40. **a** Anterior maxillary occlusal view; the x-ray tube angle is +65°. **b** Lateral maxillary occlusal view; the x-ray tube angle is +60°

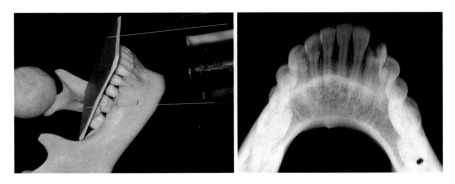

Fig. 2.41. Anterior mandibular occlusal view. The angle between the x-ray film and the tube is -55°

for anterior views, size 2 for posterior views, and for small jaws, a small size-0 film can be used. Modeling putty is useful for stabilizing the skull or mandible. Plastic or Styrofoam bite blocks are used for holding the dental film. In order to stabilize the film, white orthodontic or red rope wax can be adapted to the Styrofoam film holder and pressed onto the occlusal or incisal edges of the teeth. When it is possible to place the film within 20° of parallel to the teeth or alveolar process, a parallel technique should be used. If it is not possible, the bisecting angle technique should be employed. In both instances, the film and holder can be stabilized by adapting an appropriate thickness of wax (thin as possible) to the film holder and gently pressing it onto the incisal or occlusal surfaces of the teeth.

Occlusal techniques are simple to accomplish by adapting a small amount of orthodontic wax onto the front surface of the occlusal film and the teeth. The skull should be stabilized with modeling putty so that the film is parallel to the floor.

Bitewing radiographs can be made by articulating the skull and mandible and stabilizing them with elastics and wax. It is most convenient to stabilize the skull with modeling putty, keeping the occlusal plane parallel to the floor.

2.8.3.2
Radiography of Tooth/Bone Fragments

Often occlusal or size-2 film will be most useful for imaging fragments of tooth and bone (Fig. 2.42). The film should be placed on a flat surface and the specimen situated so that the long axis of the teeth or alveolar process is parallel to the front surface of the film. Red rope wax is used to stabilize the fragment.

2.8.3.3
Radiography of Loose Teeth

Imaging of loose teeth can be accomplished as for tooth and bone fragments (Fig. 2.43). In addition to standard images through the buccal-lingual surface,

individual teeth can be imaged through the mesial-distal surface to provide right-angle views.

The imaging of individual, loose teeth is simple and straight forward. Like photography, it will likely be desirable to make multiple exposures of the same specimen with variable exposure factors. Loose teeth can be radiographed in two planes to provide the typical mesial-distal view as well as buccal-lingual views. Ideal image geometry is simple to achieve by placing the film onto a flat surface and temporarily attaching the tooth in the desired orientation to the film using some dental rope or orthodontic wax. The size of the film will depend upon the requirements of the project. Multiple teeth will fit onto an occlusal-size radiograph and can be exposed together, or teeth can be imaged individually with smaller periapical film.

2.9
The Radiographic Appearances of Some Selected Diseases of the Teeth and Jaws

There are many texts dedicated to oral and maxillofacial pathology and radiology that can provide a detailed and comprehensive description of the many diseases that can affect the teeth and bone. The purpose of this section is to introduce the reader to the radiographic appearance of dental caries, periodontal disease, periapical disease, and pericoronal disease.

2.9.1
Dental Caries

Dental caries is a disease of the teeth that results in demineralization of the enamel and dentine and eventual cavitation of the tooth. Caries can occur on any tooth surface, but is more common on the occlusal and proximal (mesial and distal) surfaces of teeth. The smooth surfaces of teeth are more resistant to decay. Radiographically, the disease appears as areas of decreased radiographic density within the enamel and dentine. While dental caries can be detected visually

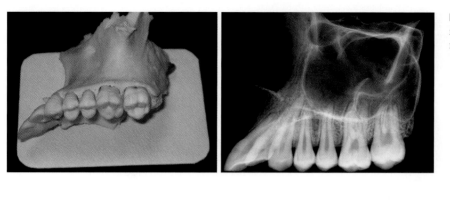

Fig. 2.42. Tooth/bone fragment stabilized with red rope wax on occlusal film

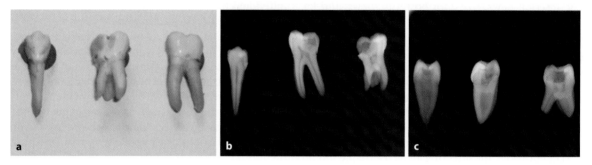

Fig. 2.43. a Loose teeth stabilized on occlusal film with small pieces of red rope wax. **b** Images of loose teeth showing typical mesial-distal projection. **c** Images of same teeth rotated 90° showing buccal-lingual projection

on most of the tooth surfaces, small lesions occurring at or just below the contact area of the mesial and distal surfaces of teeth may only be seen with bitewing radiographs. Radiographs can also demonstrate the depth of caries that are evident visually on the occlusal surface (Fig. 2.44). There are many causes for decreased radiographic density on the tooth including anatomical variations, developmental defects in the enamel (hypoplasia), wear (attrition), and fracture. Carious lesions are usually identified by tooth number and tooth surface.

2.9.2.
Periapical Inflammatory Disease

As the carious lesion advances or traumatic tooth fracture exposes the dental pulp to the oral environment, bacteria will eventually reach the dental pulp and cause pulpal inflammation and eventual pulp necrosis. The infection will eventually go beyond the tooth through the apical foramen of the roots and invade the periodontal ligament space and surrounding bone. At this point, radiographic signs will become evident indicating infection of the periapical tissues. Early signs of disease include a widening of the periodontal ligament

space around the apex of the root and a loss of definition of the lamina dura. Eventually, frank bone loss will occur, creating a radiolucent lesion of bone around the root apex. This is called a rarefying osteitis. Occasionally, in more chronic conditions, a zone of increased bone density will surround the rarefying osteitis or the widened periodontal ligament space and this is termed sclerosing osteitis (Fig. 2.45)

2.9.3.
Periodontitis

The radiographic appearance of healthy periodontal bone shows an alveolar crest that is located about 2.0 mm apical to an imaginary line that would connect the cemento-enamel junctions of adjacent teeth. The first stage of periodontal disease is gingivitis, which is inflammation of the gingival tissue in response to chronic accumulation of plaque and calculus around the teeth. If left untreated, the inflammation and infection will eventually extend to the alveolar bone causing destruction of the bone supporting the roots of the teeth. The two most common patterns of periodontal bone loss are horizontal bone loss and angular (or vertical) bone loss (Fig. 2.46).

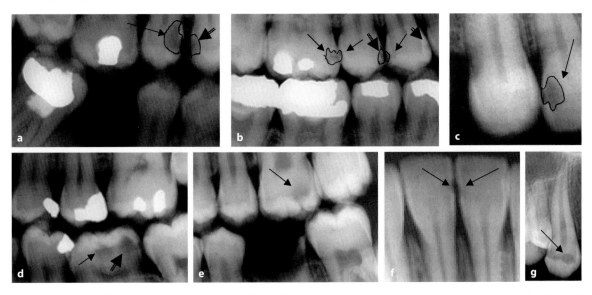

Fig. 2.44. a Advanced interproximal carious lesion at 15 mesial and early dentine lesion at 14 distal. **b** Early dentine lesion at 16 mesial, early enamel lesion at 15 distal, late enamel lesion at 15 mesial, early dentine lesion at 14 distal, and early enamel lesion at 14 mesial. **c** Late dentine lesion at 12 distal. Note the spread of carious lesion within dentine. **d** Occlusal and mesial caries at 46. **e** Occlusal caries in dentine at 26 and severe carious destruction of tooth 36. **f** Enamel caries at 31 mesial and 41 mesial. **g** Buccal or palatal smooth surface caries at 13

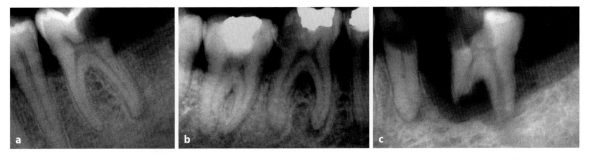

Fig. 2.45. a Periapical inflammatory bone loss (rarefying osteitis) at the distal root of the first molar. **b** Rarefying osteitis at the mesial and distal roots of the second molar and distal root of the first molar. **c** Rarefying osteitis of the second premolar and first molar. Note the root resorption at the first molar, which is occasionally seen associated with periapical inflammation

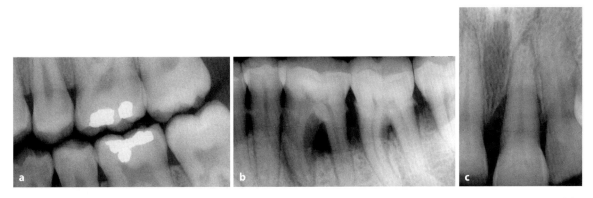

Fig. 2.46. a Appearance of normal periodontal bone. Note that the alveolar crests are within 2.0 mm of the cemento-enamel junctions. **b** Calculus deposits around the cervical area of the teeth and horizontal bone loss. Note the loss of sharp definition of the alveolar crest and bone loss between the roots of the molars (furcation involvement). **c** A vertical bone defect on the mesial surface of the left central incisor (21)

2.9.4.
Osteomyelitis

In some cases, localized bone inflammation around the roots of a necrotic tooth or periodontally involved teeth will extend more generally into the bone and bone marrow and produce an osteomyelitis. There are various forms of the disease depending on the virulence of the disease and the resistance of the host. In more acute forms, the radiographic appearance will be one of patchy bone destruction that extends beyond the teeth. In chronic cases, it is common to see a diffuse increase in bone density around the teeth (Fig. 2.47). Other radiographic features of osteomyelitis include the presence of sequesta, which appear as islands of radiopaque bone surrounded by radiolucent bands, and a typical form of periosteal reaction that looks like thin laminations of bone over a cortical surface. This finding is more commonly found in the young. It is also possible to see enlargement of the bone producing asymmetry from left to right. Osteomyelitis of the maxilla is extremely uncommon, although localized bone inflammation due to periapical inflammation and periodontitis is common.

2.9.5
Pericoronal Disease

The most common teeth to be impacted are the third molars, followed by the maxillary canines. Impacted teeth are surrounded by a uniform, thin radiolucent band called the follicle space (Fig. 2.48a). This is surrounded by a thin sclerotic zone, the follicle cortex. In situations where the tooth is partially erupted, inflammation and infection can occur around the crown; this is termed pericoronitis. Radiographically, the follicle can appear enlarged and the follicle cortex will become less apparent (Fig. 2.48b). The dentigerous cyst is a developmental cyst that can arise from the dental follicle surrounding the crown of the impacted teeth. Although the lesion is benign, it can become large and cause local destruction of bone, expansion of cortical bone, and displacement of teeth (Fig. 2.48c).

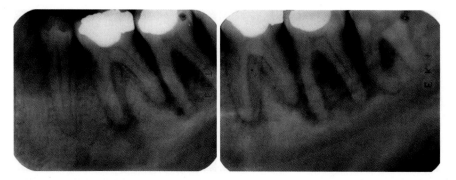

Fig. 2.47. Osteomyelitis of the mandible. Note areas of periapical and periodontal bone destruction, and below the roots of the teeth a sclerotic bone pattern that accentuates the appearance of the radiolucent inferior alveolar nerve canal. Bone sequestrum located between the roots of the second molar

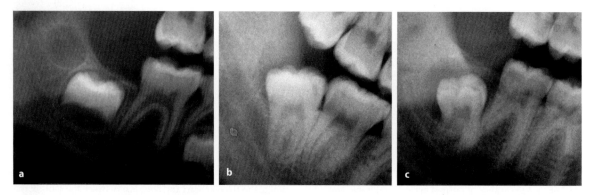

Fig. 2.48. **a** Normal dental follicle of developing second molar (note early development of third molar follicle with no tooth formation evident). **b** Partially impacted third molar with enlarged dental follicle and loss of follicle cortex typical of pericoronitis. **c** Enlarged follicle around a bony impacted third molar, indicating dentigerous cyst formation

2.10
Applications in Paleoradiology

A list of applications in paleoradiology can be found in Tables 2.5 and 2.6, and are illustrated in Figs. 2.49–2.73.

2.10.1
Three-dimensional CT in Paleoanthropology

For the last decade, three-dimensional CT has played an increasing role in the evaluation of hominid fossils (Mafart et al. 2002). This advanced technology, which combines multislice CT and imaging computer software, has allowed an innovative approach to the nondestructive study of those precious and rare materials. The range of indications for three-dimensional CT imaging includes the evaluation of skull intraos-

Table 2.5. Paleoradiology domains

1. Anatomical paleoradiology
Tool for the study of anatomy of:
skeletal remains
mummies and bog bodies
hominid fossils
2. Diagnostic paleoradiology
Tool for detecting ancient diseases (paleopathology)

Table 2.6. Paloradiology in bioarcheology

Human skeletal and teeth remains
Animal skeletal and teeth remains
Mummies: human and other animal
Bog bodies
Hominid fossils
Paleobotany
Paleontology
Soil matrix
Cremation urns
Skeletons in jars

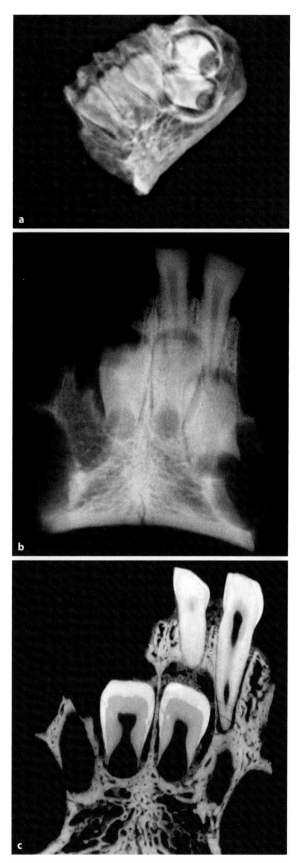

Fig. 2.49. Mandible with unerupted teeth (Prei Khmeng, Cambodia, 2000 years old). **a, b** X-rays. **c** Micro-CT image

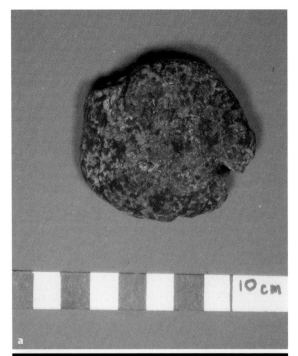

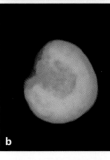

Fig. 2.51. **a** Fossilized gallstone. **b** X-ray of fossilized gallstone (Courtesy Prof Brothwell)

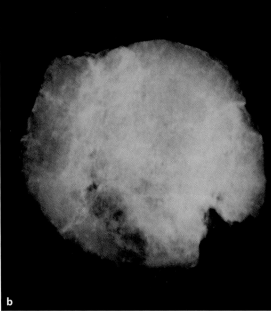

Fig. 2.50. Iron age breadroll, York (Courtesy Prof Brothwell)

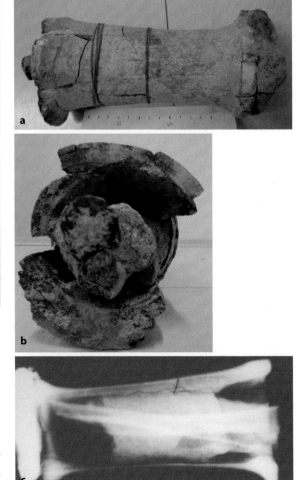

Fig. 2.52. Human forearm bone within an elephant tusk (Phum Snay, Cambodia, 2000 years old). **a, b** Specimens. **c** X-ray

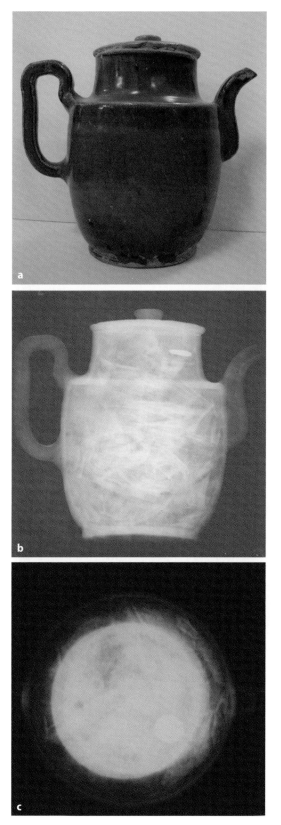

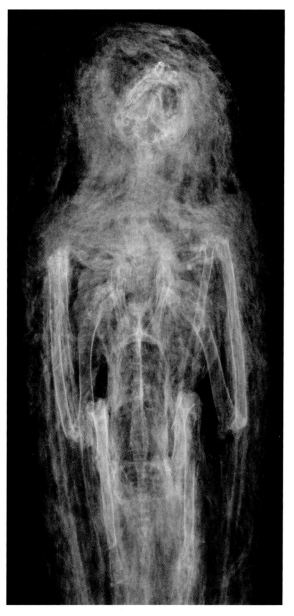

Fig. 2.54. Egyptian bird's mummy

Fig. 2.53. a–c Cremated bones within an urn (APSARA-Siem-reap, Cambodia 16th century). **a** Specimen. **b, c** X-ray

Fig. 2.55. Egyptian bird's mummy containing eggs instead of a bird!

Fig. 2.56. a Fossilized tree bark. **b** X-rays of fossilized tree bark shown in **a**

Fig. 2.57. Skeleton with bone matrix. **a** Specimen. **b** CT of tibia within soil matrix. **c** Three-dimensional CT of the hip (Prei Khmeng, Cambodia, 2000 years old)

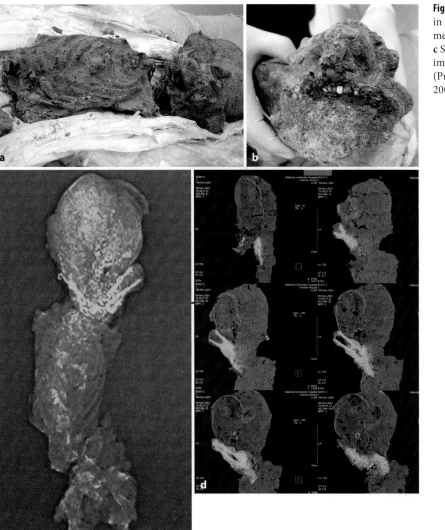

Fig. 2.58. Skeleton of a child in soil matrix. **a** Specimen. **b** X-ray of specimen. **c** Skull specimen. **d** CT images of the skull specimen (Prei Khmeng, Cambodia, 2000 years old)

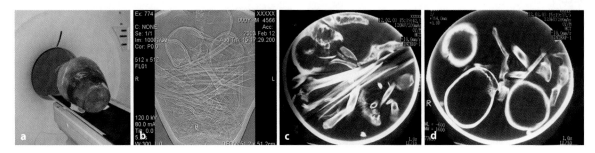

Fig. 2.59. Bones in a jar. **a** Specimen. **b** Scout view. **c**, **d** CT images (Cardamoms, Cambodia, 16th century)

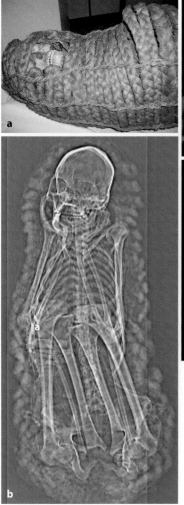

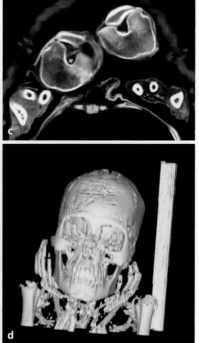

Fig. 2.60. Peruvian mummy. a Scout view. b X-ray. c CT of the knees. d Three-dimensional CT of the head and shoulders (Courtesy of Professor Villari)

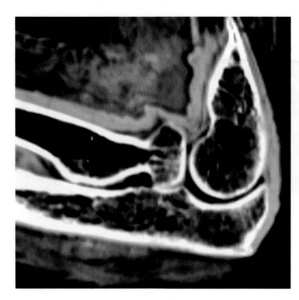

Fig. 2.61. CT of an elbow from an Egyptian Mummy

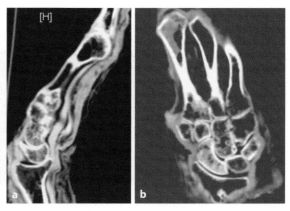

Fig. 2.62. Wrist of an Egyptian mummy. a Sagittal CT image. b Coronal CT image

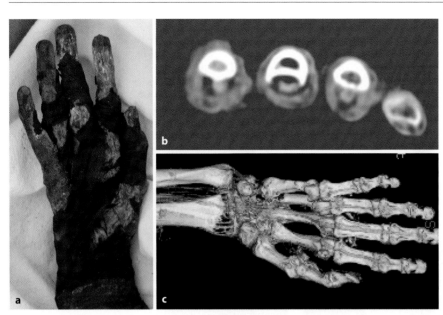

Fig. 2.63. Hand-wrist tendons. **a** Specimen. **b** Axial CT scan. **c** Three-dimensional CT of the flexor tendons

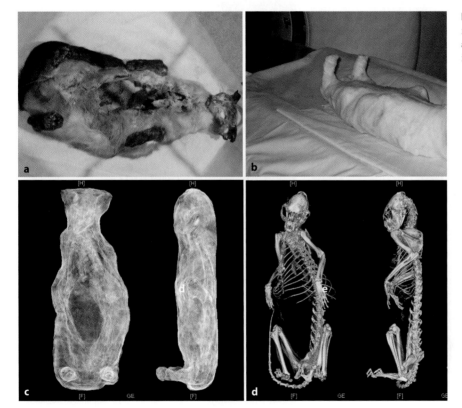

Fig. 2.64. Experimental mummification of a cat. **a, b**: Specimen. **c, d**: CT images

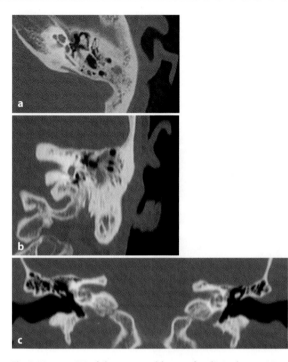

Fig. 2.65. a–c CT of the temporal bone of a clinical case. Normal anatomy

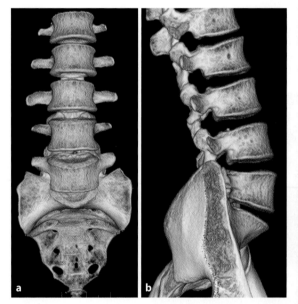

Fig. 2.66. a, b Three-dimensional CT of a normal lumbar spine (courtesy of General Electric)

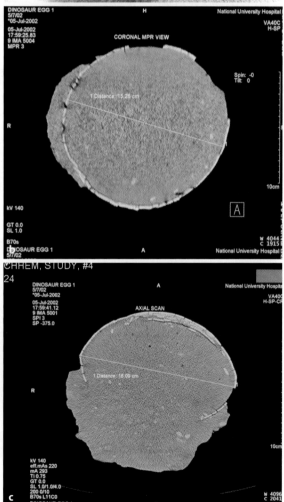

Fig. 2.67. Cretaceous dinosaur's hatched egg, from China (courtesy of Dr. Yap). **a** Specimen. **b, c** CT images

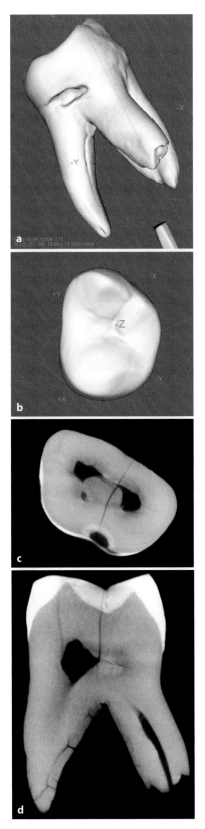

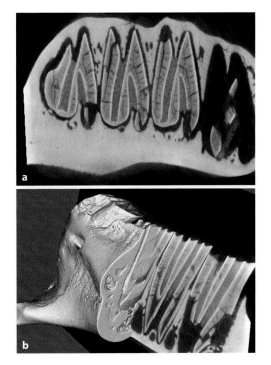

Fig. 2.69. Micro-CT of a fossilized guinea pig. **a** Two-dimensional micro-CT. **b** Three-dimensional micro-CT

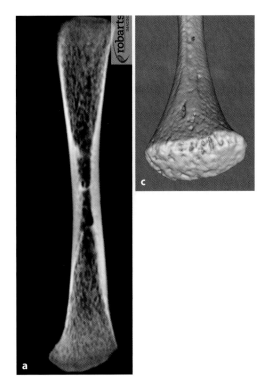

Fig. 2.68 a–d. Micro-CT of a tooth (Cardamoms, Cambodia 16th century)

Fig. 2.70. Infant humerus. **a** Two-dimensional micro-CT. **b** Three-dimensional micro CT

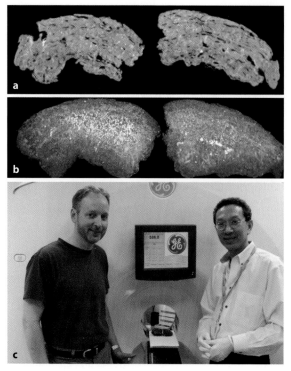

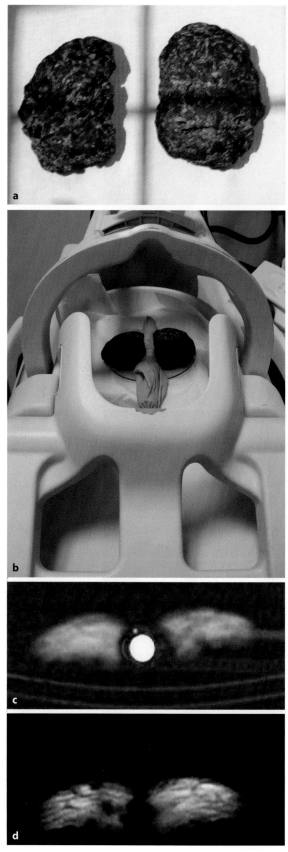

Fig. 2.71. A 3200-year-old Egyptian natural mummy's brain. **a** Specimen (Courtesy of the Royal Ontario Museum). **b** Micro-CT. **c** Professor Holdsworth (*left*) and Professor Chhem (*right*)

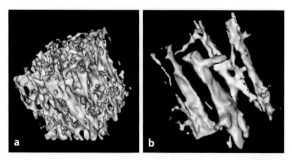

Fig. 2.72. a MicroCT of trabeculae in normal vertebral body. **b** MicroCT of trabeculae that are rare, thick and vertical in vertebral body with hemangioma (Specimen Courtesy of Dr El Molto)

seous cavities such as sinuses and the semicircular canals of the inner ear, as well as morphometric analysis (Table 2.7). This technology helps in the creation of a virtual endocast of hominid brains, separating it from its rock matrix without inducing any damage to the original specimen. The use of imaging digital data leads to a virtual representation of fossils that al-

$\longrightarrow$

Fig. 2.73. Egyptian natural mummy's brain (3200 years old). **a**, **b** Specimen (Courtesy of Royal Ontario Museum). **c**, **d** MRI

Table 2.7. Three-dimensional (3D) CT imaging in paleoanthropology

Intraosseous cavities of skulls such as sinus and semi circular canals of the inner ear
Post-cranial skeletons
Morphometric analysis
Virtual endocast of hominid brains
Virtual representation of fossils
Study of ontogeny, phylogeny and diagenesis

Table 2.8. Paleoradiology in mummy studies

Presence of a human or animal skeleton
Forgeries
Age of the mummy
Mummification process
Paleopathology

Table 2.9. CT and mummies: indications

Whole-body (+3D CT-surface rendering)	
Museum display	
Gross anatomy/ Anthropological	
Media industry = public education and entertainment	
Database	
Limited scanning time:	Royal Mummies
	Frozen Mummies
	Others
Imaging software R & D	
Region of interest (two-dimensional, multiple planes)	
Hypothesis-driven research	
Paleopathology[a]	

[a]Do not use 3D process because it will defeat the purpose of the CT scan, which was to detect lesions inside the bone. The 3D technique will hide lesions underneath the bone surface

lows the study of ontogeny (individual development), phylogeny (speciation), and diagenesis (fossilization). Radiological studies of archeological soil have been used to evaluate the stratigraphic boundaries and their precise locations, to detect signs of disturbance, to highlight mixing and redeposition, and variation in sediment variation of density. This technique had been proven useful in the analysis of paleoenvironment in order to elucidate the depositional and post-depositional context of samples, and finally to guide subsampling selection. There is no doubt that CT would add tremendous information and precision to x-ray techniques of imaging archeological soils in the near future (Zollikofer and Ponce de Leon 2005).

2.10.2 CT and Burials

When the bones have been cremated or treated then preserved in jars, x-ray study helps in the analysis of the content of those burial urns. The ritual storage of skeletons in large jars is widespread among prehistoric populations, especially in Southeast Asia. The jars are usually preserved in remote areas of the jungle or in mountain caves. The content of the jars can be easily studied by visual inspection. However, CT scan will help in documenting the relative positions of each bone within those jars, as the digital data collected will remain a precious spatial document of bone distribution within the container, even after the bones have been removed for anthropological study. The precise position of those individual bones may be of great value to cultural anthropologists interested in the study of mortuary rituals in the future. CT is also useful for the study of skeletons in their soil matrix without any need to remove the soil (Chhem et al. 2004). Finally, x-ray and CT have been used for archeological soil and sediment profile analysis (Butler 1992).

2.10.3 CT and Mummies

CT study of mummies, especially those from Egypt, is extremely popular among both scholars and the lay public (Hoffman et al. 2002). CT has been used for the evaluation of the content of the mummies, without the need to unwrap them. The cross-sectional images and three-dimensional capability of the new generation of multislice CT scans have allowed the creation of not only mummies' images in two and three dimensions, but also facial reconstructions, which are extremely appealing for museum display. Beyond this main goal aimed at lay public education, CT had also been used in many scientific studies of mummies, including evaluation of the skeleton and dry tissues, the age of the mummy, the mummification process, the identification of burial goods, and a few paleopathological studies (Tables 2.8 and 2.9). Among these studies, the role of CT in the detection ancient diseases is yet to be validated. A review of the scientific literature on CT evaluation of mummies that has been published since the first CT of Egyptian mummy was performed by Dr. Harwood-Nash in Toronto in September 1976 (Harwood-Nash 1979) has shown a methodology that needs much refinement, especially in terms of accuracy in diagnosing ancient skeletal and dental pathologies.

References

Ballinger PW (1982) Merrill's Atlas of Radiographic Positions and Radiologic Procedures, 5th edition. Mosby, St. Louis

Batchelar DL, Cunningham IA (2002) Material-specific analysis using coherent-scatter imaging. Med Phys 29:1651–1660

Bushong SC (2004) Radiologic Science for Technologists, 8th edition. Elsiver Mosby, St. Louis

Butler S (1992) X-radiograpahy of archaeological soil and sediment profiles. J Archaeol Sci 19:151–161

Chhem RK (2006) Paleoradiology: imaging disease in mummies and ancient skeletons. Skeletal Radiol 35:803–804

Chhem RK, Ruhli FJ (2004) Paleoradiology: current status and future challenges. Can Assoc Radiol J 55:198–199

Chhem RK, Venkatesh SK, Wang SC, Wong KM, Ruhli FJ, Siew EP, Latinis K, Pottier C (2004) Multislice computed tomography of two 2000-year-old skeletons in a soil matrix from Angkor, Cambodia. Can Assoc Radiol J 55:235–241

De Backer AI, Mortele KJ, De Keulenaer BL (2004) Picture archiving and communication system – Part one: Filmless radiology and distance radiology. JBR-BTR 87:234–241

Gallet J, Titus H (2005) CR/DR systems: what each technology offers today; what is expected for the future. Radiol Manage 27:30–36

Harwood-Nash DC (1979) Computed tomography of ancient Egyptian mummies. J Comput Assist Tomogr 3:768–773

Hoffman H, Torres WE, Ernst RD (2002) Paleoradiology: advanced CT in the evaluation of nine Egyptian mummies. Radiographics 22:377–385

Hohenstein P (2004) X-ray imaging for palaeontology. Br J Radiol 77:420–425

Holdsworth DW, Thorton MW (2002) Micro-CT in small animal and specimen imaging. Trends Biotech 20:s1–s6

Johns HE, Cunningham JR (1984) The Physics of Radiology, 4th edition. Charles C. Thomas, Springfield, IL

Lang J, Middleton A (1997) Radiography of Cultural Material. Butterworth Heineman, Oxford

Lauterbur PC (1973) Image formation by induced local interactions: examples employing nuclear magnetic resonance. Nature 242:190–191

Mafart B, Delingette H, Gerard Subsol; International Union of Prehistoric and Protohistoric Sciences (2002) Three-dimensional Imaging in Paleoanthropology and Prehistoric Archaeology. British Archaeology Reports. Archaeopress, Oxford

McErlain DD, Chhem RK, Bohay RN, Holdsworth DW (2004) Micro-computed tomography of a 500-year-old tooth: technical note. 55:242–245

Nishimura DG (1996) Principles of Magnetic Resonance Imaging. Stanford University Press, Palo Alto, CA

Recheis W, Weber GW, Schafer K, Knapp R, Seidler H, zur Nedden D (1999) Virtual reality and anthropology. Eur J Radiol 31:88–96

Schueler BA (1998) Clinical applications of basic X-ray physics principles. Radiographics 18:731–744

Zollikofer C P, Ponce de Leon MS (2005) Virtual Reconstruction: A Primer in Computer-Assisted Paleontology and Biomedicine. John Wiley Sons, New York

CHAPTER 3

The Taphonomic Process, Biological Variation, and X-ray Studies

3

DON R. BROTHWELL

3.1
X-raying the Whole Range of Bioarcheological Materials

While calcified tissue is rightly considered to most benefit from radiographic study, it is important to keep in mind that other biological materials may be explored using this technique. Indeed, the potential for its application to the study of ancient organic remains has yet to be fully realized. For this reason, it is appropriate to review here, as broadly as possible, the different kinds of bioarcheological materials that can potentially be radiographed. Having said that, it should be acknowledged that the advent of fine digital imaging, and especially computed tomography (CT) scanning, may transform the prospects for studying small organisms and smaller pieces of bioarcheological material.

Clearly the three main divisions of biology that need to be considered are plant remains, invertebrates, and vertebrates. These are not of equal importance in terms of their variety and preservation at archeological sites, but they all deserve some consideration nevertheless. It is also essential to include a consideration of taphonomic and diagenetic factors, as the chances of preservation and the degree of preservation can vary considerably from site to site (Lyman 1984, 1994). There are very few simple rules when it comes to the preservation of bioarcheological remains. Variation in alkalinity and acidity are not well correlated with the degree of preservation. Microradiography has been one means of assessing the histological integrity or changes occurring in buried bones (van Wagenen and Asling 1958). Alkaline calcareous "petrification" of soft tissue in cave sediments and acidic preservation in peat bogs demonstrate the need to be prepared to accept many different states of decay. Muds and silts may penetrate deep into plant remains and bones, so that the radiographic image may be very different indeed to what it is like in life. Natural mummification (drying) may not preserve all kinds of tissue to the same degree, so the x-ray images may again be distorted from the real-life appearance. Ritual mummification of the kind best known from

Egypt may change tissues in other ways, especially with the application of such substances as natron (hydrated sodium carbonate from natural evaporite deposits), resins, and bitumen. The hardened tissues may produce unusual radiographic shadows, and in the case of the intervertebral discs, cause changes that look remarkably like alkaptonuric arthritis.

In the case of wood, marine wood borers can cause a complex of tunnels into which silt may drift, and on land, woodworm, and death-watch beetles can transform ancient timber. Similarly, bone can be seriously damaged and changed by insects, and for instance at least one species of bee can destroy bone to the extent that the ragged holes simulate malignant metastatic deposits. Rodents are also well known for damaging bone, and circular damage produced by them has even been misidentified as human surgical intervention. Finally, there are the natural processes affecting bone and other tissues. Localized natural erosions may look remarkably like osteolytic activity, while some degree of mineral replacement (as part of a gradual fossilization process) can result in an apparently increased bone density.

All of these matters have to be considered when applying radiological techniques to bioarcheological material (Douglas and Williamson 1972, 1975; Ely 1980 Morgan 1988; Østergard 1980). As yet, perhaps the most neglected aspect of x-raying in archeology and anthropology is at the field level. This is as yet totally neglected, and perhaps this is because most situations demand action in the laboratory and not on site. But there are situations where portable radiography, of the kind used by veterinary specialists, could be an advantage at an excavation. For instance, it could have been useful after the discovery of the Lindow bog body, to have inserted film under the small exposed area of the body in the bog, in order to try and see the extent and orientation of the body still in the peat. This would have needed "trial and error" radiographing, as the density of the body and the peat are very similar. In special situations, radiography on site might assist in determining what a deposit contained. From my own experience, I know that it would have been useful to x-ray a "slice" of midden deposit at an Orkney site, where fin rays were still in correct relationships, whereas attempts to excavate them resulted in their disintegration. Similarly, plant remains within fine water-borne deposits can be usefully x-rayed before attempts are made to clear and remove the fragile structures.

3.2
The Evaluation of Botanical Remains

From the publication evidence, it would seem that few botanists or archeologists consider the radiographic techniques have any value to their disciplines. It can certainly be said that in comparison with its clear value in studying vertebrate remains, x-rays have more limited application to plant remains. But nevertheless, there are a variety of situations where it has proved useful to identify plant material, consider structural aspects, or on the evidence of x-rays to consider conservation measures.

Unless soft plant remains have been "invaded" by other elements that assist in their preservation, it is likely that only the harder and more durable plant tissues will mainly remain in archeology. Taphonomic factors may obscure detail or differentially erode or damage plant remains. In the case of marine timber, these can be quite rapidly attacked by "wood-boring" organisms, including the infamous shipworm (Teredo, a bivalve mollusc). The shipworm settles on wood as very small larvae, initially creating only a tiny hole in the timber surface. They then grow and excavate into the wood, producing a large series of internal tunnels. Fortunately for radiography, the tunnels are lined with calcium carbonate, which shows up relatively well in x-rays (Fig. 3.1). Not all ancient marine wrecks suffer damage from wood-destroying organisms, and the 17th-century Swedish warship, the Wasa, appears to have escaped serious decay, possibly because of the microenvironment in Stockholm harbor, where it capsized. However, as a conservation measure, it is clearly a good policy to x-ray samples of timbers and other wood objects from marine sites as a check and precautionary measure.

Waterlogged sites in particular have produced a variety of timbers used in different ways, and selected ones have been of considerable value in building up an important dendrochronological database. While sections can often be cut (Fig. 3.2) in such wood samples, CT scans provide a precise and nondestructive means of counting growth rings (Kuniholm 2001; Tout et al. 1979). It could also provide accurate information on fluctuations in growth-ring thickness, linked to environmental factors over time.

A very different kind of tubular structure of plant origin was found at a Roman site in York (UK). Depo-

Fig. 3.1. X-ray of an experimental wood sample, showing damage from marine wood borers (simulating ancient shipworm destruction). Moss x-ray 1. Courtesy of Karla Graham and English Heritage

sits on this site were rich in iron and the result was numerous ferruginous tubular structures, up to 5–10 cm in length. But these were not Roman objects, as suspected, but when x-rayed were seen to be rotted tree roots that had attracted iron oxide from the surrounding soil. Archeological objects of different kinds can benefit from radiographic evaluation of the botanical material. This may produce very unexpected finds, as in the x-ray of the Neolithic "Iceman's" calf leather belt pouch, which revealed tinder fungus, flint tools, and a bone awl (Figs. 3.3 and 3.4). Similarly, an x-ray of the retoucheur revealed a worked antler point driven deep into the medullary canal of the limewood grip. In addition, CT scanning revealed the contents of the quiver – 14 arrow shafts, in this fur bag, stiffened by an attached hazel rod (Egg et al. 1953) (Fig. 3.5).

Fig. 3.2. Radiograph of growth lines in medieval waterlogged wood from York

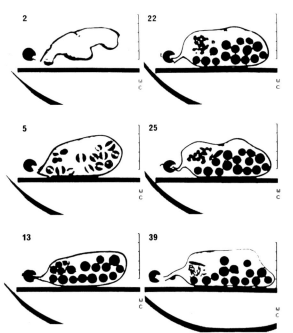

Fig. 3.5. Computed tomography (CT) scan revealing the contents of Iceman's quiver. Courtesy K. Spindler

Fig. 3.3. The Iceman's belt pouch of calf leather. Courtesy K. Spindler

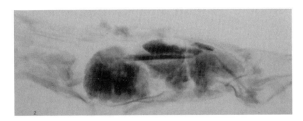

Fig. 3.4. Contents of the belt pouch, as shown by x-ray, including fungus, bone awl, and flint tool. Courtesy of K. Spindler

With the advent of digitally enhanced x-rays, it is now possible to enlarge and obtain much better detail of ancient flax fabrics, for instance. Indeed, the finer structures of any ancient objects of material culture made from plants, from fish traps to chariot wheels, can potentially be further revealed by x-raying. Without deconstructing the object, radiographs allow a closer evaluation of the carpentry in wooden objects, hidden tenons, dowelling, and so forth. Sediment deposits and corrosion, under normal visual examination, can obscure the detail that x-rays can once more reveal.

The study of ancient waterlogged lacquerware provides a good example of how xeroradiography has been applied successfully in the past to the study of archeological objects. In the case of a series of early Chinese finds, they presented various conservation problems, although they appeared visually to be in good condition (Jackson and Watson 1995). There was no sign of fungal colonization, but the water content varied considerably, as did the lacquer surfaces. Low-energy x-rays were used, with exposures around 60 kV, 3 mA for 10 s. Investigation by xeroradiography was found to be ideal. Under-surface detail, including repairs to the objects, were clearly revealed. This technique is clearly of great value in the study of lacquerware (Fig. 3.6).

The other category of plant material that deserves radiographic consideration is food remains. In particular, the internal detail of poorly defined or consolidated cereal or other plant deposits, bread, or even coprolites, could potentially be revealed by x-ray (Figs. 3.7 and 3.8). Breads may still reveal cereal fragments or inorganic inclusions, while sediment-covered or consolidated cereal grain or other stored food might yield more information from radiographs. Also, some coprolites are rich in coarse, but mastica-

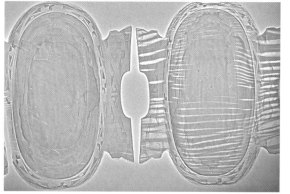

Fig. 3.6. Xeroradiographic image of ancient Chinese lacquerware, with internal detail clearly visible. Courtesy of Jacqui Watson, English Heritage

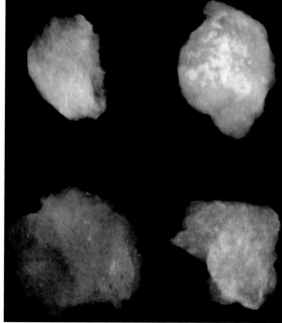

Fig. 3.8. Ancient coprolites, containing some contrasting structures

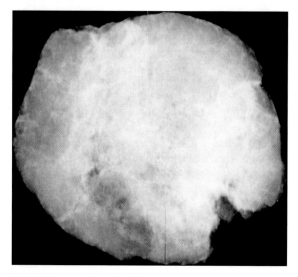

Fig. 3.7. Iron Age bread roll (burnt), x-rayed to show the inner granular detail

ted plant debris, and x-rays provide an ideal method of scanning such material before any specimens are selected for further laboratory analysis.

3.3
Radiological Aspects of Zooarcheology

Paleontologists have made use of radiographic techniques more enthusiastically than zooarcheologists in the past. Indeed, in some instances, it has been a highly successful tool in revealing the less commonly seen finer anatomy of organisms. For instance, from fine shale of the Ordovician and Devonian periods, the pyrite-preserved soft parts of trilobites within the shale were revealed by x-ray (Robison and Kaesler 1987). An ideal x-ray of a fossil demands a balance

between sufficient penetration through the fossil and matrix (needing high voltage) and sufficient light/dark contrasts (needing a low kV and a long exposure time). Depending on the lithology, fossil hydroids, graptolites, fish, and other vertebrate species have been revealed successfully by this technique (Harbersetzer 1994; Longbottom 2005). Heavily fossilized skeletal material can present special problems with regard to obtaining sufficient detail of the internal structures, especially if the interior of the bone is invaded by siliceous deposits (Fig. 3.9).

Although it is mainly vertebrate remains that have received radiographic study in archeology, it should be mentioned that there is clearly potential when considering invertebrate remains, for the extraction of information by means of x-rays. There is certainly a need for colleagues specifically working with invertebrates to consider the potential value of radiographic techniques, and perhaps undertake experimental investigations in order to evaluate what might be achieved with some kinds of archeological material. For instance, would it be possible, by means of digital x-rays, to detect not only insect damage, but also the actual animals buried within other organic remains? For instance, insect damage to bone might be confirmed radiographically by detecting the insect remains deep within the calcified tissue. However, experiments would be needed to establish that insect structures could be identified within bone or other tissue. Similarly, insect damage to mummified remains is

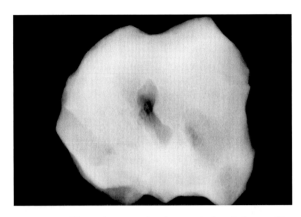

Fig. 3.9 Fossil bone from South Africa with detail obscured by siliceous deposits

not usually confirmed by the internal detection of the invertebrate species.

By far the most important group are the molluscs, the shipworm already being mentioned. As yet, there has been little use of x-rays in the study of the internal structure of shells, and nothing in archeology, although the study of shell growth lines by CT scanning could be achieved without sectioning. The application of radiography to the analysis of vertebrate remains is, however, a very different story.

Although calcified tissues preserve relatively well on archeological sites, and detailed studies on these bones and teeth have been undertaken for many years, the radiographic investigation of these finds have been surprisingly neglected. This is in part due to the general unavailability of x-ray machines, but more particularly because radiography has not been seen as a worthwhile research tool. However, the situation is changing and departments of archeology, museums, and major archeological trusts, increasingly have access to x-ray facilities on a day-to-day basis.

In the past, some of us have had to make use of a portable x-ray machine, of the kind used in veterinary work. This has the advantage of being movable to museum stores or even into field conditions (if there is an energy source; Clutton-Brock et al. 1980). Setting up a portable machine within buildings is likely to result in some concern if the protection screening does not totally contain any radiation scatter. It is not enough for the person operating the machine to wear a lead rubber protection apron. The ideal, in our experience, is to have available a lead rubber sheet for below the specimens to be x-rayed, as well as a clamp-held lead rubber sheet (fixed in a circular or squared shape) as a vertical surround. The x-ray head would then be directed down centrally into this protected area, producing minimal scatter.

However, if an x-ray machine can remain in a fixed laboratory position, and the usual range of objects

for x-ray are no longer than a human femur, then the Faxitron cabinet x-ray system is a compact and ideal model. The machine is versatile – with an ideal kV range of 10–120 kV and 3 mA – and can not only take x-rays of biological materials in general (including mummified tissue and fabrics), but also of soil cores, ceramics, and metals.

In establishing an x-ray machine for the first time, and if a digital facility has not been made available, it is of course necessary to consider the purchase of developing chemicals and a range of accessory equipment. These will include a viewing illuminator, cassettes, processing tanks, and frames. Safe lights are not recommended, since (unlike photographic film) x-ray film should be developed in complete darkness. Medical x-ray film is not recommended (unless live animals are being x-rayed) because the control is poor and details can be blurred. Industrex AX and SR films from Kodak give far superior results, particularly the finer-grained SR film, which can be used in slide mounts to project clear images of fish vertebrae and mummified tissue for measurement and diagnostic purposes. Positioning aids should include radiotranslucent pads of cotton wool and plastic foam, for example.

3.4
Positioning and Image

As in all radiography (Douglas and Williamson 1972, 1975), in x-raying bioarcheological materials, we aspire to obtain a good portrayal of the structures being examined. Unlike the radiography of modern biological specimens, however, we have the challenging task of differentiating between antemortem biological reality and the taphonomic and burial factors that can result in a range of visual artifacts. Burial can result in changes in contrast that may demand some experimenting with exposure factors. There may also be postmortem erosions or internal silting (through areas of damage or nutrient foramina), which can simulate extra ossification (Fig. 3.9). On the other hand, calcified tissue can be perfectly preserved, as shown by the micro-CT scanning of a prehistoric guinea pig mandible (Fig. 3.10).

The positioning of skeletal material depends upon the relevance of the x-rays to a particular investigation (Figs. 3.11 and 3.12). If, for instance, the x-rays are to explore cranial sinus variation, or the cortical tissue thickness of a long bone, or the standard cranial measurements of dried or mummified dogs, then every effort should be made to position the specimens in standard lateral or anteroposterior (or other) orientations. If, however, the x-ray is to investigate specific abnormality or pathology, it may well be necessary to position each specimen uniquely in order to bring out the maximum detail. In the case of a complex structure such as the skull, it is important to remember that the superimposition of bone contours from the left and right halves can confuse, if not conceal, detail that the x-ray is in fact intended to highlight. For example, possible healed injury to the zygomatic arch and maxilla may be revealed by a dorsoventral x-ray (Fig. 3.13), but not by a simple lateral x-ray. However, an angular positioning of the cranium, relative to the x-ray film, might bring out further detail of the trauma. Whether employing standard x-ray positions (lateral, dorsoventral, anteroposterior), or otherwise, it is important to become familiar with radiographic skeletal anatomy before attempting interpretations of abnormality. Building up an archive of normal and pathological bone radiographs of different species is invaluable in this respect. It is also useful to tabulate, for reference purposes, the kV and timing needed to produce good x-rays of various skeletal materials.

3.5
Taphonomic Aspects of Bones and Teeth

The complex nature of the interaction between the variables that help to prevent or assist the preservation of vertebrate remains after death is now well described (Lyman 1984, 1994). Thus taphonomy needs only a brief consideration in relation to the radiography of skeletal remains. The reason for consider-

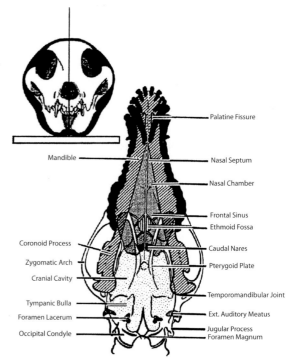

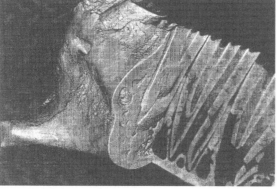

Fig. 3.10. Micro-CT scan of a guinea pig mandible from Colombia (supplied by Professor Chhem)

Fig. 3.12. Diagram of the dog skull in dorsoventral orientation, and the x-ray detail revealed in this position. (Modified from Douglas and Williamson, 1972.)

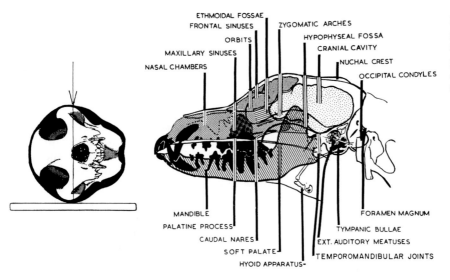

Fig. 3.11 The need for radiographic positioning, as seen for example in the lateral positioning of the dog skull. After Douglas and Williamson (1972)

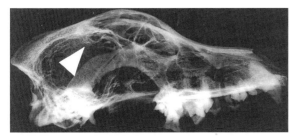

Fig. 3.13. Slightly angled positioning of an iron Age dog skull, to explore by x-ray an area of healed damage to the right orbital area (arrowed)

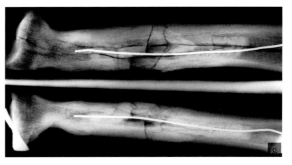

Fig. 3.14. Relatively good detail of a restored fossil cave-bear radius

ing this matter at all is because site influences on the preservation of bones and teeth are highly variable and have an effect on the quality of the radiograph obtained on archeological specimens. The most ideal environments, ensuring better than usual preservation of vertebrate remains, are found in arid or cold environments. Frozen mammoths and other Pleistocene species recovered from sites in Alaska and Siberia are examples of how well preserved prehistoric remains can be.

In temperate climates, such as in northern Europe, and in tropical environments, bones and teeth may become eroded, demineralized (becoming radiotranslucent), or show varying degrees of mineral replacement (becoming radio-opaque). Nevertheless, following some experimentation in terms of kVs and exposure timing, the most problematic material is likely to yield to radiography. For instance, the human remains from Broken Hill (Kabwe), Zambia, are impregnated with lead and zinc ores, but can still be penetrated by x-rays. Well-fossilized bones from the Tor Newton cave in the south-west of England give satisfactory results, as exemplified by the broken and restored cave-bear radius (Fig. 3.14). Ironically, in my experience, a pathological bone from Iron Age Danebury presented more of a radiographic problem. The bone appears to have been fractured, with massive callus formation. But this whole area has been impregnated with inorganic sediment, which has greatly reduced the contrasts within the specimen (Fig. 3.15).

Generally, bones and teeth, whether recent or subfossil, need kVs and exposure times within a relatively narrow range of values. Smaller mammals, small fish, and birds usually have far thinner bones, and this needs to be taken into account when deciding on the kV and timing. In the case of birds, there are some exceptions to this rule, for the large flightless species can display a very robust build, especially in the long bones of the legs (Fig. 3.16).

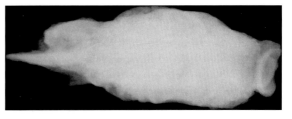

Fig. 3.15. Deformed Iron Age long-bone fragment, with x-ray detail obscured by impregnated sediments

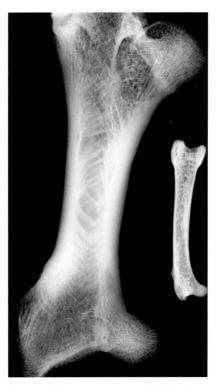

Fig. 3.16. The robust structure of a Moa femur, compared to the fragile nature of a chicken bone

3.6
Measurement from X-rays

As yet, very few radiographic studies have been undertaken on nonhuman vertebrate species, with a view to comparing different population samples. There is certainly a need to begin to establish standards for within-group means and variation of the kind, for instance, revealed for the human hand (Clauser 1962) or vertebral column (Todd and Pyle 1928). Variation in dry human bones have also been explored, both in terms of cortical thickness and sinus size and shape (Brothwell et al. 1968). In other mammals, x-rays could assist in the measurement of cranial angles, for instance (Grigson 1975).

In the study of Holocene mammals, Horowitz and Smith (1990, 1991) provide good examples of the ways in which the radiography and measurement of normal bones may assist in the full evaluation of change through time. By x-raying a series of caprine metacarpals, for instance, and taking comparable cortical thickness measurements of goat and sheep through the Holocene, they were able to demonstrate a secular trend in the reduction of cortical bone thickness. In further studies, these researchers were able to demonstrate cortical thickness changes at Jericho and other Near Eastern sites, suggesting secular changes after the Neolithic and perhaps especially through the Bronze Age. Factors such as intensive milking in females and size selection in males could contribute to this variation. From a methodological point of view, it is important to keep in mind that in thicker long bones to be measured, the time and kV needed to reveal the inner contour of the cortical bone may fade out the outer alignment of the bone. For this reason, the outer bone surface in the region where any measurements are to be taken should be marked with lead crayon or with a taped lead strip (or even pins). An alternative could be to support the bone on foam pads, simulating soft tissue.

As well as variation in bone thickness, there is clearly other variation that has been poorly explored so far. How much variation (indicative of regional, gender, or age differences) is there in the extent or nature of trabecular tissue at the articular ends of bones? Similarly, what variation is to be seen in the extent or complexity of sinus systems (especially in terms of sexual dimorphism, maturity, or regional-variety differences)?

In the case of chickens, it would be valuable to have far more information, revealed by radiography, of the inner aspects of long bones of the legs, indicating calcium storage (as medullary bone) before egg laying begins (Fig. 3.17). By means of CT scans, it is of course possible to reconstruct whole skeletons on screen, where no comparative dry bone specimens are

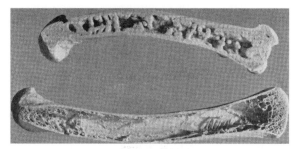

Fig. 3.17. Bone sections of chicken femora, which could be studied nondestructively by x-ray. The laying animal (left) has considerable amounts of internal medullary bone. Courtesy Dr. K. Simkiss

available. An example of this can be seen in the CT reconstruction of a turtle skeleton, required in order to study ancient turtle remains (Frazier 2005).

Further measurement related to radiographic variation in bones is concerned with the recording of ancient bone density differences, developed in the past two decades (Lyman 1984). CT scanning has been shown to be ideally suited to the investigation of a range of archeological materials, from fossils to more recent faunal specimens (Lam et al. 1998). Bone density variation can be used to consider taphonomic questions, such as whether species representation is influenced by bone density and its affect on differential survival (Lam et al. 1999, 2003). Bone density studies of this kind are further needed in relation to age-group composition within species and to fluctuating environmental stress.

3.7
X-raying Aspects of Growth

While the determination of age from the skeleton and dentition may be determined from "external" details such as bone size, degree of maturity and union of epiphyses, or the extent of the eruption of attrition of the teeth, there are circumstances where radiography may provide valuable additional information. It need hardly be mentioned that developing teeth within the jaws can be readily revealed by lateral radiographs (Fig. 3.18). Similarly, epiphyses that are only partially united to the diaphysis may be assessed with regard to the degree of closure by recourse to x-rays. Projecting radiographs of archeological fish vertebrae onto a screen can make it easier to count the growth rings and thus age the individual. Digital x-rays are even better for this task.

While the radiographic ageing of human adults has now received some study and review (Sorg et al. 1989), there is a need for far more detailed studies

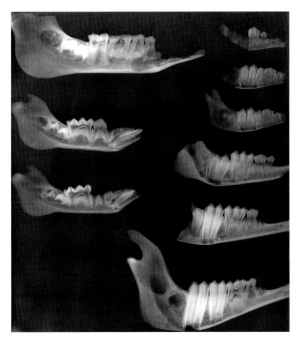

Fig. 3.18. X-rays of immature pig (left) and sheep jaws (right), showing variable internal age-related detail of tooth development

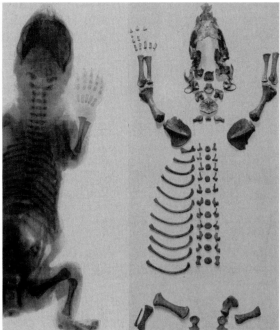

Fig. 3.19. Neonate cave-bear bones (right) compared with the x-ray of a modern neonate brown bear (left; after Abel and Kyrle 1931.)

of a similar kind on other mammals. As it is easier to x-ray living populations of domesticates, it is to be hoped that studies will eventually be initiated on possible age-related change, as mentioned previously. We also need growth information on wild species, as exemplified by a comparison of a neonate cave bear, with the x-ray of a neonate brown bear (Abel and Kyrle 1931) (Fig. 3.19).

One might hope that in the future, the interests of the zooarcheologist are combined with the biologist in order that research designs may include questions perhaps of special relevance to investigating past populations. Having said that, the radiographic study of skeletal growth provided by Whenham, Adam, and Moir (Wenham et al. 1986) on development in fetal red deer provides precise data of the kind that we need more of in archeology. Mention of this study serves to emphasize this continual need to relate modern data to the resolution of problems in the past, and indeed, there are now studies on a growing range of species, from Rhesus monkeys to horses and cattle (Brown et al. 1960; MacCallum et al. 1978; van Wagenen and Asling 1958; Wenham et al. 1969, 1986). With this kind of information for reference, ancient material can be aged more accurately (Kratochvil et al. 1988). It might also be noted here that studies on living species may have relevance to the interpretation of abnormal growth in the past (Rudali 1968).

A final point regarding growth is that it is clearly related to the degree of sexual dimorphism. Part of the final growth differentials leading to sexual dimorphism in parts of the skeleton may in fact result in changes to the internal architecture of bones, and this matter has yet to be assessed.

3.8
Frozen, Dried, and Mummified Bodies

For well over two centuries the arid conditions of Egypt, Peru, and other regions of the world have produced bodies of interest to archeology. Somewhat unusually, even part of the second woolly rhinoceros found frozen at Starunia in Poland was x-rayed for detail of the foot (Fig. 3.20) (Novak et al. 1930). Some of the bodies, representing humans and other vertebrate species, have naturally dried after burial. Others, especially many from Egypt, have been submitted to mummification procedures – usually by the application of natron. After removing various inner organs and embalming all the soft tissues, the Ancient Egyptians wrapped the bodies of kings, queens, and special dignitaries, as well as a variety of ritual animals (cats, birds, shrews, even baby crocodiles and snakes).

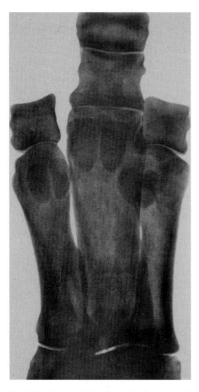

Fig. 3.20. X-ray of the foot of a frozen and well-preserved woolly rhinoceros from Poland. After Novak and colleagues (1930)

Soon after the discovery by Röntgen late in 1895, there was world interest in the potential use of x-rays (Holland 1896 – quoted by David 1978), both in clinical work and other forms of examination. As early as 1896, the fine detail of a frog skeleton had been described (Bétoulières 1961) and Dr. Thurston Holland of Liverpool investigated several bundles, including a bird. It is also recorded that in May 1897, Dr. Charles Leonard, using an x-ray machine built at the University of Pennsylvania (USA), was able to obtain radiographs of a Mochica body from Pachacamac, Peru (see Fiori and Nunzi 1995). Very soon afterwards, Flinders Petrie published x-rays of a Fifth Dynasty Egyptian body from Deshasheh (Petrie 1898). The bones displayed evidence of Harris lines, bone pathology whose significance was not recognized for some decades after. The radiographs were remarkably good, considering the rudimentary nature of these first x-ray machines and the slow-exposure glass photographic plates that were used.

Peru has continued to produce the dried bodies of various species, especially human, and to a lesser extent dogs, guinea pigs, camelids, and certain other species. In the case of some bodies, the dried tissue and hair can at times obscure bone detail, but generally the conformation of the bones and teeth is good,

and both size and shape variation can be studied from such material.

In early Egypt the preservation of ritual animals could, on occasion be as good as for humans, (although some of the x-rayed ibis and birds of prey from Saqqara, excavated in 1994, appear to have been allowed to rot before mummification) but, as in the case of human mummies, the external wrappings are detrimental to the actual study of the vertebrate remains – except by radiography. The investigation of all such mummies by x-rays is a worthwhile goal for the future, and some nonhumans are at last being studied in this way (Ikram and Iskander 2002). Positioning of mummy bundles is a methodological problem that remains to be resolved. The difficulty is knowing how best to position the bundles in relation to the x-ray film, in order to get the best views of the inner skeleton (in terms of identifying species, noting pathology, and possibly taking measurements for comparative purposes). The nature of the problem is illustrated by the Egyptian bird mummy from the Petrie Collection at University College London. The body, which appears to be damaged in the middle, also appears to be headless. But the head and neck are curved onto the body between the (long) wings (Figs. 3.21 and 3.22).

At times, the radiography of mummy bundles reveals special surprises. For example, during an examination of bird mummy bundles at UCL, one specimen turned out to be three eggs, even though it was wrapped to look like an adult bird (Fig. 3.23). A supposed cat mummy has been revealed by x-ray to be a fake, and was simply a large nonfeline bone made up to look like a cat mummy (Pahl 1968). The linen wrappings on such Egyptian mummified bundles have been found to be virtually radiotranslucent, even when covered in resin, and can be ignored when setting the time and kV. However, the presence of sand and mud with the body is a common problem and can obscure parts of the skeleton. Therefore, frontal and profile radiographs should be taken as a matter of course. Given good detail, species identification is far more possible, as for instance in the case of a fish mummy identified on bone evidence as Nile cat fish (*Eutropius niloticus*; Leek 1976).

Others who have x-rayed animal mummies have had surprises. In one instance, what appeared to be a crocodile from the external appearance, in fact turned out to be a collection of four crocodile skulls arranged one after the other to simulate a complete body. Another miniature bundle was shown by radiography to be a small gerbil (David 1978), while a "dog" mummy turned out to be two sections of a human long bone. The coffins constructed for animal mummies are usually in the shape of the animals preserved, but the actual body size within the coffin may turn out to be a poor fit (even allowing for shrinkage with dehydrati-

Fig. 3.21. X-ray of a naturally dried recent bird, to show how well the skeletal structures can appear

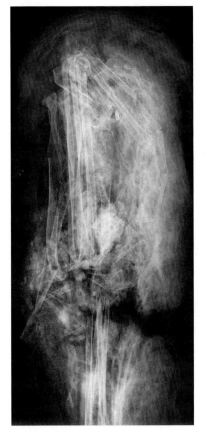

Fig. 3.22. X-ray of an Egyptian bird mummy, where interpretation is made difficult owing to body positioning for mummification

Fig. 3.23. X-ray of a bird mummy (as apparent from the external wrappings) that was in fact three eggs

on). For instance, the Late Period young cat (No. 9303) in Manchester Museum (UK) turned out on x-ray to be much smaller than the adult-sized coffin (Fleming et al. 1980). As well as the contents, museums may also be surprised at how much damage can be suffered by the skeleton within a mummified package through repeated handling and transportation. Even though the external surface appears undamaged, the bones inside are very fragile and do not "give" as easily as the linen wrappings. An x-ray record should be kept of such items to check on the degree of deterioration; this is especially important if the package is dissected at a later date. Records need to be kept of measurements of animals seen in x-ray. By measurement, it has been shown that a series of young Nile crocodiles displayed similar body sizes, suggesting that these immature animals were kept and used at specific sizes for ritual purposes (Owen 2001).

3.9 Microradiography

Although interest in magnified radiographs extends back to the turn of the last century, the 1950s saw the most significant advances, which then extended over the next four decades. Fine-grained emulsions contributed to these developments, as well as to the eventual development of special microfocal x-ray tubes.

While Britain made a significant contribution to this field of radiography, it should be acknowledged that the potential application of microradiography to archeological studies was also being explored elsewhere. Osteon appearances were, for instance, being revealed in prehistoric human bone from a variety of periods and sites (Magdalenien, Neolithic, Bronze Age, Hallstatt, and so on). The magnification of ×40 provided very good evidence of osteon sizes and remodeling, as well as postmortem cracks and erosions (Baud and Morganthaler 1956). The evaluation of the histology of calcified tissues and of diagenetic processes could clearly be assisted by such procedures, but there were also other possibilities.

Using equipment developed in particular by Buckland-Wright (1976, 1980, 1989) and based on initial designs by Cosslett and Nixon (1952), a range of bone and tooth specimens was studied. In each case, there was a clear advantage in obtaining a much enlarged image. The specimens included sections of mammal bones as well as skeletal elements from small mammals and amphibians. In the bone sections, patterns of remodeling during growth could be clearly seen. The investigation of small mammal bones was related to my interest in resolving the problem of the differential identification of species from individual bones (and the possibility that activity-related differences might be revealed in the internal architecture of some bones).

The final reason for exploring microradiographic techniques was in order to consider their value in detecting small-scale pathology. This has been mentioned previously (Baker and Brothwell 1980), and does allow vertebrate paleopathology to extend to small species. It obviously permitted small-scale pathological lesions to be viewed in much better detail even within larger bones (e.g., microfractures in osteoporotic trabecular bone). Small metastatic deposits, for instance, could not only be viewed in enlarged form, but considered more precisely in terms of raggedness of contour and shape. Joel Blondiaux and colleagues (Blondiaux et al. 1994) demonstrated the value of microradiography in the study of the concentric remodeling of bones changing shape as a result of leprosy. All of this is now being replaced by the advent of digital radiography, which is proving to be an excellent technique.

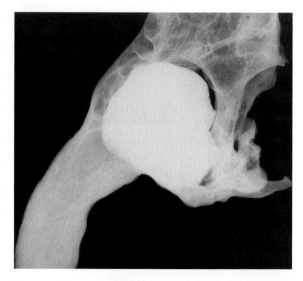

Fig. 3.24. Abnormal horncore of a sheep, displaying narrowing as a result of "thumbprints"

3.10
Problems of Differential Diagnosis

The importance of radiographs in the correct interpretation of animal pathology needs little emphasis (Andrews 1985; Kold 1986; Morgan 1988). However, it would seem useful in this chapter to review a few cases in order to provide evidence of the varying extent to which x-rays may assist in answering diagnostic and research problems. Pathology will be expanded on in Chapter 6.

3.10.1
Horncore "Thumbprints"

It is well known to those working on horncores, especially of sheep, that abnormal narrowing may occur along regions of the horncore, giving depressions reminiscent of thumbprints pressed into clay. As yet, their etiology has not been finally resolved, although the problem has been investigated in several zooarcheology laboratories (Albarella 1995). In x-ray (Fig. 3.24), there is no evidence of internal crushing damage to the bone tissue and there is no deflection of the frontal sinus extending into the horncore. In this case, it does not therefore provide any clues to the etiology of the condition. On a technical point, the white mass in Fig. 3.24 is plasticine, used to support the skull in the optimal position for the radiography of the horncore. Whenever possible, props of this kind should be of a material that does not show up on x-rays. Alternatively, for some bone positioning, such props may be marginal to the x-ray and can

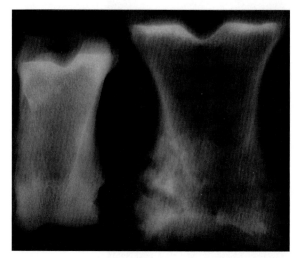

Fig. 3.25 The quality of x-rays in two specimens that originally formed part of the quick scanning of a relatively large number of fragments

thus be cut out of the final illustration for publication.

3.10.2
Leg Bones

The two specimens shown in Fig. 3.25 formed part of a group of six specimens placed on a Kodak Industrex AX ready pack film, as part of a quick scanning procedure for a series of bone pathology from the Southampton University (UK) faunal unit. In the case of a dog tibia with a bowed shaft, the question was whether this anomaly was the result of genetics, rickets or trauma. Bone remodeling is minor, but extends to the medullary surface, and there are no obvious changes to the articular ends or Harris lines. In view of these points, it seems likely that the pathology results from a greenstick fracture rather than a state of malnutrition. In the case of the sheep metopodial, there are two aspects to consider. Within the marrow cavity are lighter masses extending along half of the shaft. Had there been a single mass of this kind in the shaft of a human long bone, the differential diagnosis would have had to include a possible infarction, but these multiple masses within the sheep bone are quite certainly the result of inorganic silty material being intruded into the bone through cracks or foramina. What is pathological, and becomes clearer in x-ray, is a swelling near the mid-shaft. This is situated on only one side, and consists of cortical bone expanding internally into the medullary cavity, as well as externally. Within this expansion is an area of less dense bone. The most likely explanation of this pathology seems to be that the animal suffered a restricted injury to the leg, resulting in moderate infection, with bone damage and necrosis in the region of the trauma. Healing occurred, resulting in new bone formation surrounding the original damaged area.

3.10.3
Vertebrae

The two pig vertebrae shown next in x-ray (Fig. 3.26) are also from early Southampton. Additional radiographic views were taken of these and other vertebrae, but only one displayed marked pathology. While the upper vertebra is a normal bone for comparison, the other visibly displays a marked irregularity to the vertebral body. In x-ray, there is clearly deep variation in the degree of bone density, confirming that this is a severe osteomyelitis.

3.10.4
Significant Bone Loss

Pathological processes may result in significant bone loss, either locally or throughout the skeleton. In our own species, postmenopausal osteoporosis immediately comes to mind, but other species may show various kinds of bone loss. In fact, severe osteoporosis has been noted in a sheep from Portchester (Fig. 3.27) and might be the result of bad over-winter conditions followed by pregnancy and then death.

In contrast to the previous specimens, the femur from the leg of the second Lindow bog body (Brot-

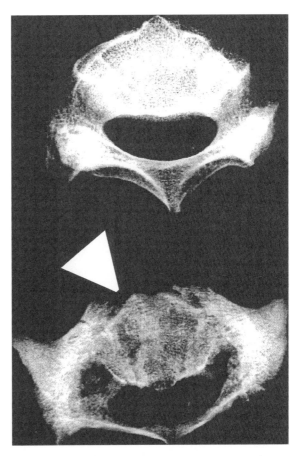

Fig. 3.26. Photocopy of a radiograph of two Saxon Southampton pig vertebrae, one being normal and the other displaying osteomyelitis of the vertebral body (arrowed)

Fig. 3.27. X-ray of a sheep femur from Portchester, England, with severe osteoporotic bone loss and cortical thinning

hwell and Bourke 1995) displays considerable bone loss and reduced density as a result of post-mortem changes while buried in the acid peat (Fig. 3.28). The acidity of deposits is clearly a factor that should not be forgotten when considering variation in bone density, while burial on chalk can result in a deeply etched surface. In the case of bones that are demineralized to any extent, pressure on them during burial can cause noticeable distortions in shape. For instance, wooden stakes pressed into the Danish Huldremose bog body caused severe deformity of a forearm (Fig. 3.29) and a femur (Brothwell et al. 1990). Skulls can deform like deflated footballs.

3.10.5
Abnormal Cavities in Bone

Abnormal cavities within bone can develop for various reasons, and their interpretation will be influenced by their position in relation to the area of skeleton or dentition. They may not always be obvious on archeological bone since postmortem erosion can mask an osteolytic lesion. Radiographs can show whether or not a sclerotic lining exists, thus indicating a pathological condition that would otherwise have been missed. If there is the possibility that there is apical infection (an abscess) at one or more tooth positions – even without clear external evidence – it is advisable to radiograph the jaw. Cysts, neoplastic processes and trauma with infection may all lead to considerable bone destruction and remodeling. In an Iron Age pig mandible from Danebury (Fig. 3.30), there is considerable bone destruction posterior to the canine, which in lateral view can be seen to extend under part of the posterior dentition. In the case of a Saxon phalanx from Southampton, an infection was obvious at the proximal joint, but only in x-ray can

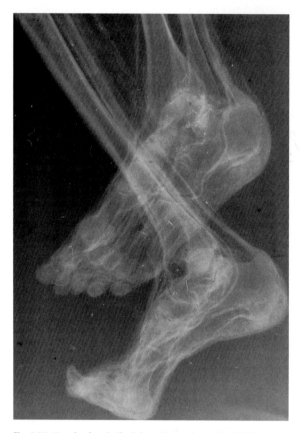

Fig. 3.28. Partly decalcified foot bones from the Huldremose bog body, showing poor bone density

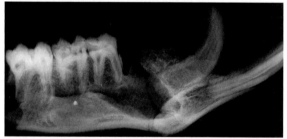

Fig. 3.30. Radiographic detail of a Danebury Iron Age pig, showing severe inflammatory changes within the jaw

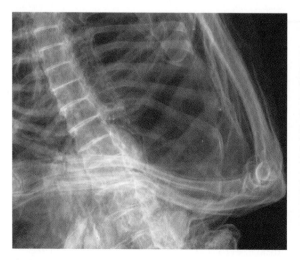

Fig. 3.29. X-ray of the thorax area of the Danish Huldremose bog body, displaying postmortem changes of the forearm

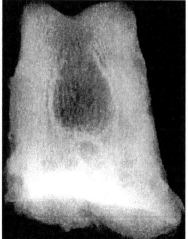

Fig. 3.31. Photocopy of a radiograph of a Saxon phalanx with a well-defined internal abscess

the large abscess cavity be seen deep within the bone (Fig. 3.31).

3.11
Conclusion

In concluding this chapter, it is obvious that radiographs greatly enhance the prospects for a more reliable and full evaluation of ancient plant and animal remains. Clearly not all bioarcheological remains can be studied by radiography, and selection must be undertaken. The nature of the material and the archeological problems posed may indicate what specimens are especially worthy of x-raying. It is worth emphasizing that it is often possible to scan quite a number of specimens arranged close together, as seen in Fig. 3.32. This is clearly demonstrated by Gejvàll

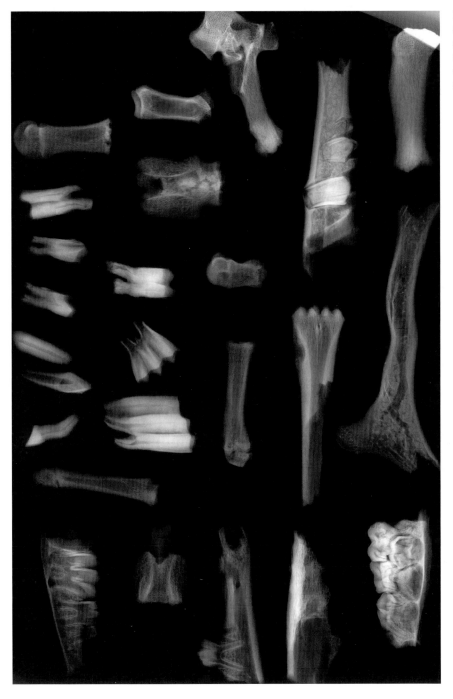

Fig. 3.32. An example of the radiographic scanning of a series of bone fragments, carried out to confirm or otherwise any possible abnormalities

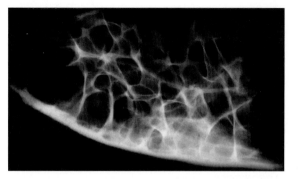

Fig. 3.33. Posterior occipital fragment in x-ray of an ancient cow skull, which displays on the surface some antimortem holes; etiology as yet unknown

(1969) in assembling a selection of animal paleopathology from the prehistoric site of Lerna in Greece. Costs are always a matter for consideration, and multiple-specimen x-rays are far more economical and quite sufficient for the basic scanning of significant numbers of specimens.

It should be mentioned that zooarcheologists can see dry-bone pathology that extends beyond the normal range of cases normally seen by veterinary colleagues on the living. In such instances, it is especially important to have x-rays available for discussion and tentative diagnosis. For instance, there are quite a number of cases of ancient bovid skulls displaying on the external surface of the occipital bone (at the posterior nuchal aspect) one, or often more than one, rounded perforation (Brothwell et al. 1996), but as yet only preliminary radiographic studies have been undertaken (Fig. 3.33). Yet, the chances of a correct diagnosis rests on assessing the external apertures in relation to the connecting internal sinus complex, which means in turn the need to "close in" on the perforations by digital radiography or local CT scanning.

Exotic species are not to be excluded when considering species representation and animal health, for even ancient societies kept tamed or caged wild forms. As a result of poor feeding, conditions such as rickets and nutritional secondary hyperparathyroidism can occur (Porter 1986).

Finally, it should be noted that even with apparently "normal" bone it is advisable to routinely x-ray important specimens before subjecting them to any destructive tests, in order to reveal any hidden pathologies and provide a record for future reference.

3.12
Summary

While in the past, radiography has been neglected by archeological botanists and zooarcheologists, there is now clear evidence of its potential in assisting in the evaluation of basic structures, of growth and aging, normal variation, and in the complex field of vertebrate paleopathology. There will be little excuse in the future for ignoring this technique in the field of bioarcheology. In particular, it appears clear that digital radiography and CT scanning may transform the quality of information derived from such investigations.

Acknowledgments

I would like to express my considerable appreciation to Naomi Mott, who worked with me at University College London (UK) on aspects of x-raying both the skeletal and mummified remains of various species of vertebrate, and Jacqui Watson for considerable help and advice on botanical aspects.

References

Abel O, Kyrle G (1931) Die drachenhöhle bei Mixnitz. Speläol Monogr 7–9:1–953

Albarella U (1995) Depressions on sheep horncores. J Archaeol Sci 22:699–704

Andrews AH (1985) Osteodystrophia fibrosa in goats. Vet Annu 25:226–230

Baker J, Brothwell D (1980) Animal diseases in archaeology, Academic Press, London

Baud C-A, Morganthaler PW (1956) Recherches Sue le Degré de minéralisation de l'os humin fossile par la méthode microradiographique. Arch Suisses d'Anthropol Gén 21:79–86

Bétoulières P (1961) Les débuts de la radiologie à Montpellier. Monspel Hippocrat 13:23–28

Blondiaux J, Duvette J-F, Vatteoni S, Eisenberg L (1994) Microradiographs of leprosy from an osteoarchaeological context. Int J Osteoarchaeol 4:13–20

Brothwell DR, Bourke JB (1995) The human remains from Lindow Moss 1987–8. In: Turner RC, Scaife RG (eds) Bog Bodies, New Discoveries and New Perspectives. British Museum Press, London, pp 52–58

Brothwell DR, Molleson T, Metreweli C (1968) Radiological aspects of normal variation in earlier skeletons: an exploratory study. In: Brothwell DR (ed) The Skeletal Biology of Earlier Human Populations. Pergamon, London, pp 149–179

Brothwell DR, Liversage D, Gottlieb B (1990) Radiographic and forensic aspects of the female Huldremose body. J Dan Archaeol 9:157–178

Brothwell D R, Dobney K, Ervynck A (1996) On the causes of perforations in archaeological domestic cattle skulls. Int J Osteoarchaeol 6:472–487

Brown WAB, Christofferson PV, Massler M, Weiss MB (1960) Postnatal tooth development in cattle. Am J Vet Res 21:7–34

Buckland-Wright JC (1976) The micro-focal x-ray unit and it application to bio-medical research. Experientia 32:1613–1615

Buckland-Wright JC (1980) Qualitative and quantitative assessment of tissue organization in normal and diseased organs. In: Ely RV (ed) Microfocal Radiography. Academic Press, London, pp 147–196

Buckland-Wright JC (1989) A new high-definition microfocal x-ray unit. Br J Radiol 62:201–208

Clauser CE (1962) X-ray anthropometry of the hand. Technical Report No. AMRL-TDR-111, Wright Patterson Air Force Base, Ohio

Clutton-Brock J, Dennis Bryan K, Armitage P, Jewell P A (1990) Osteology of the Soay sheep. Bull Br Mus Nat Hist (Zool) 56:1–56

Cosslett VE, Nixon WC (1952) An experimental x-ray shadow microscope. Proc R Soc Ser B, 140:422–431

David R (1978) Mysteries of the Mummies. Book Club Associates, London

Douglas SW, Williamson HD (1972) Principles of veterinary radiography. Bailliere Tindall, London

Douglas SW, Williamson HD (1975) Veterinary radiological interpretation. Heinemann, London

Egg M, Goedecker-Ciolek R, Waateringe WG-V, Spindler K (1993) Die gletschermumie vom ende der steinzeit aus den Ötztaler Alpen Jahrbuch des Römisch-Germanischen Zentral-museums 39 Mainz

Ely RV (1980) The evolution of microfocal x-ray equipment. In: Ely RV (ed) Microfocal radiography. Academic Press, London, pp 1–41

Fiori MG, Nunzi MG (1995) The earliest documented applications of x-rays to examination of mummified remains and archaeological materials. J R Soc Med 88:67–69

Fleming S, Fishman B, O'Connor D, Silverman D (1980) The Egyptian Mummy, Secrets and Science. Handbook 1. University Museum, Philadelphia

Frazier J (2005) Marine turtles – the ultimate tool kit: a review of worked bones of marine turtles. In: Luik H, Choyke AM, Batey CE, Lougas L (eds) From Hooves to Horns, From Mollusc to Mammoth. ICAZ, Tallin, pp 359–382

Gejvall N-G (1969) The Fauna: Volume 1, Lerna, a Preclassical Site in the Argolid. American School of Classical Studies at Athens, Princeton

Grigson C (1975) The craniology and relationships of four species of Bos: II, basic craniology, Bastaurus L. Proportions and angles. J Archaeol Sci 2:109–128

Harbersetzer J (1994) Radiography of fossils. In: Leiggi P, May PJ (eds) Vertebrate Palaeontological Techniques, Vol 1. Cambridge University Press, Cambridge, pp 329–339

Horwitz LK, Smith P (1990) A radiographic study of the extent of variation in cortical bone thickness in Soay sheep. J Archaeol Sci 17:655–664

Horwitz LK, Smith P (1991) A study of diachronic change in bone mass of sheep and goats. Jericho (Tel-Es-Sultan) Archaeozoologia 4:29–38

Ikram S, Iskander N (2002) Catalogue general of Egyptian antiquities in the Cairo Museum: non-human mummies. Supreme Council of Antiquities Press, Cairo

Jackson T, Watson J (1995) Conservation of waterlogged Chinese lacquerware from the warring states period 475–221 BC. The Conservator 19:45–51

Kold SE (1986) The incidence and treatment of bone cysts in the equine stifle joint. Vet Annu 26:187–194

Kratochvil Z, Cerveny C, Stinglova H, Lukas J (1988) Determining age of medieval cattle by x-ray examination of metapodia. Pamatky Archaeol 79:456–461

Kuniholm PI (2001) Dendrochronology and other applications of tree-ring studies in archaeology. In: Brothwell DR, Pollard AM (eds) Handbook of Archaeological Sciences. Wiley, Chichester, pp 35–46

Lam YM, Chen X, Marean CW, Frey CJ (1998) Bone density and long bone representation in archaeological faunas: comparing results from CT and photon densitometry. J Archaeol Sci 25:559–570

Lam YM, Chen X, Pearson OM (1999) Intertaxonomic variability in patterns of bone density and the differential representation of bovid, cervid, and equid elements in the archaeological record. Amer Ant 64:343–362

Lam YM, Pearson OM, Marean CW, Chen X (2003) Bone density studies in zooarchaeology. J Archaeol Sci 30:1701–1708

Leek FF (1976) An ancient Egyptian mummified fish. J Egypt Archaeol 62:131–33

Longbottom A (2005) The use of x-rays in palaeontology. Set in Stone. NHM (Palaeo) Newsletter 3:8–10

Lyman RL (1984) Bone density and differential survivorship of fossil classes. J Anthropol Archaeol 3:259–299

Lyman RL (1994) Vertebrate taphonomy. Cambridge University Press, Cambridge

MacCallum FJ, Brown MP, Goyal HO (1978) An assessment of ossification and radiological interpretation in limbs of growing horses. Br Vet J 134:366–374

Morgan J P (1988) Radiology of skeletal disease, principles of diagnosis in the dog. Iowa state University Press, Ames

Novak J, Panow E, Tokarski J, Szafer W, Stach J (1930) The second woolly rhinoceros (Coelodonta antiquitatis Blum) from Starunia, Poland. Bull Int Acad Polon Sci Lett 3:1–47

Østergard M (1980) X-ray diffractometer investigations of bones from domestic and wild animals. Amer Ant 45:59–63

Owen LM (2001) A radiographic study of thirty-nine animal mummies from Ancient Egypt. B.Sc. dissertation, University of York, UK

Pahl WM (1986) Radiography of an Egyptian "cat mummy", an example of the decadence of the animal worship in the late dynasties. Ossa 12:133–140

Petrie WMF (1898) Deshasheh. Memoir of the Egypt Exploration Fund, London

Piepenbrink H, Schutkowski H (1987) Decomposition of skeletal remains in dry desert soil. A roentgenological study. Hum Evol 2:481–491

Porter FJ (1986) Radiology of exotic species. Vet Annu 26:372–378

Robison RA, Kaesler RL (1987) Phylum Arthropoda. In: Boardman RS, Cheetham AH, Rowell AJ (eds) Fossil Invertebrates. Blackwell, London, pp 205–269

Rudali G (1968) Experimental production of hyperostosis frontalis interna in mice. Israel J Med Sci 4:1230–1235

Sorg MH, Andrews RP, Iscan MY (1989) Radiographic aging of the adult. In: Iscan MY (ed) Age Markers in the Human Skeleton. Thomas, Springfield, IL, pp 169–193

Todd TW, Pyle SI (1928) A quantitative study of the vertebral column by direct and roentgenoscopic methods. Am J Phys Anthropol 12:321–338

Tout RE, Gilboy WB, Clark AJ (1979) The use of computerised x-ray tomography for the non-destructive examination of archaeological objects. Proceedings of the 18th International Symposium on Archaeometry and Archaeological Prospection, 14–17 March 1978, Bonn, pp 608–616

Wagenen van G, Asling CW (1958) Roentgenographic estimation of bone age in the Rhesus monkey (Macaca mulatta). Am J Anat 103:163–186

Wenham G, McDonald I, Elsley FW (1969) A radiographic study of the development of the skeleton of the foetal pig. J Agric Sci Camb 72:123–130

Wenham G, Adam CL, Moir CE (1986) A radiographic study of skeletal growth and development in fetal red deer. Br Vet J 142:336–349

CHAPTER 4

Diagnostic Paleoradiology for Paleopathologists

4

RETHY K. CHHEM, GEORGE SAAB, and DON R. BROTHWELL

4.1
Introduction

4.1.1
Diagnostic Paleoradiology

Paleopathology is the study of ancient disease processes in skeletal remains using a spectrum of methods consisting of gross observation and radiological, paleohistopathological, biochemical, isotope, and DNA studies. Each of these tests carries both advantages and limitations, and almost all require the irreversible destruction of the specimen. In contrast, x-ray study is an appealing option because it can be performed without any significant damage to the specimen. According to Ragsdale, "Radiographs can be thought of as gross photographs of whole lesions and contribute much toward the orthopedic pathologist's goal of understanding the origin, structure, and mechanism of skeletal diseases", and "Maceration of residual gross material produces bone specimens that match radiographs precisely and are relevant to other fields, for example paleopathology..." (Chhem 2006; Ragsdale 1993). This underlines the important role of specimen radiological studies in both clinical and paleopathological investigations. Whether the radiation causes any alteration of the genetic materials of the specimen has yet to be documented scientifically. The accurate detection of ancient skeletal lesions faces several challenges, one of which is the lack of any clinical history, upon which the diagnostic radiologist customarily relies. An additional difficulty in establishing the final diagnosis is the impossibility of having any laboratory tests performed, such as blood, urine, or any other body fluid tests.

Of all of the medical tests available to the clinician, x-rays are the most appropriate first-line procedure for the diagnostic approach to skeletal lesions. X-ray study may bring a wealth of information on bone and joint diseases, by allowing the "visualization" of the internal structure of the bones, without the inevitable alteration and/or destruction of the specimen. Despite their nondestructive properties, radiological studies are unfortunately still underutilized in the evaluation of ancient bones.

Obtaining a good-quality x-ray study of a dry skeletal specimen is technically easy. The main limitation lies in the interpretation of the radiological findings. The fundamental problem here is to identify the most qualified scientist and expert to carry out this radiological evaluation. The professionals who perhaps most meet the required criteria are skeletal radiologists, as their medical background equips them with the scientific and medical skills to diagnose diseases of the bones and joints using x-rays. However, a skeletal radiologist with no working knowledge of physical anthropology or bioarcheology may face difficulties in interpreting diseases in ancient skeletal specimens. Skeletal radiologists probably will need to learn about the taphonomic and diagenesis processes of bone and teeth in order to avoid false diagnoses of diseases that can be explained by postmortem alterations. Radiologists must take note of the bioarcheological data of the x-rayed specimens, which include the age at death, gender, stature, ethnic group or human population, geographic location, and archeological context. Close collaboration with an anthropologist and familiarity with the main issues related to paleopathology would certainly enhance the skeletal radiologist's skills in establishing bone and joint diseases with confidence and high accuracy.

The nature of ancient skeletal remains makes them quite different from the skeleton in a live patient, and these differences will alter the accuracy and validity of any diagnostic test. The establishment of accurate terminology is the key to a valid radiological approach to the differential diagnosis (also called a gamut) of skeletal lesions. In medicine, the differential diagnosis means starting with the displayed signs and symptoms, and then trying to differentiate between the potential diseases or conditions associated with those signs and symptoms to find the correct diagnosis. The availability of valid and accurate terminology not only enhances the quality of paleopathological diagnosis, but also facilitates communication between experts from diverse backgrounds, who are working in bioarcheology. The basic key radiological patterns of the skeletal system already exist and have been used extensively and accurately in clinical situations, and most can certainly be applied to paleoradiological studies. However, some basic radiological patterns are not applicable to dry specimens. For example, joint effusion and soft-tissue abnormalities no longer exist in dry specimens, depriving the skeletal radiologist of crucial clues that would ordinarily help them to establish a diagnosis in live specimens. Obviously, standard clinical radiology is not a perfect approach to paleoradiology and paleopathology; however, diagnostic radiological methods represent one of the few "gold standards" available and are superb tools in the diagnosis of bone and joint diseases.

Despite the current development of medical imaging technology, no technique (such as, for example, computed tomography – CT, plain films, and magnetic resonance imaging – MRI) will ever establish itself as a perfect gold standard, simply because some of the human tissues associated with live bones, such as bone marrow, tendons, and muscles, are desiccated or have disappeared due to the process of taphonomy. Another fundamental fact is that current radiological procedures are extremely sensitive in detecting bone and joint abnormalities, but they are not always specific. "Pathognomonic" patterns are rare and only apply to a few skeletal lesions, including fracture, nonossifying fibroma and osteochondroma (radiologists "Aunt Minnie diagnostic"), but in the vast majority of cases it is wise to offer a differential diagnosis that includes two to three lesions because most skeletal lesions share common radiological patterns. Finally, clinical radiologists propose their differential diagnosis in light of the clinical history they have been provided along with the knowledge of patient age. Often, the two to three differential diagnoses radiologists offer are the result of a process initiated by a detailed analysis of basic radiological patterns, followed by a cognitive synthesis of what the radiological patterns combined could represent, and finally narrowed down to the most likely diagnoses. The possibility of pseudopathology, a well-known concept in paleopathology, must be emphasized and considered systematically if one wishes to avoid diagnostic error.

In conclusion, because of the nature of dry specimens, a precise diagnosis is often impossible. In some cases, paleoradiologists may establish a diagnosis that includes broad categories of skeletal diseases rather than offering a specific diagnosis, which even in clinical situations is impossible to achieve without the use of additional tests such as biopsy with histological and/or bacteriological studies.

This chapter presents a logical approach to the diagnosis of skeletal lesions using plain radiography, and offers some general basic principles on the x-ray interpretation of bone lesions for bioarcheologists with no medical or radiological backgrounds. The gamut approach is a well-known diagnostic method used by radiologists in proposing a differential diagnosis for pathologic lesions as detected by x-ray study (Chapman and Nakielny 2003; Resnick and Kransdorf 2005). This rigorous method, which is fully validated in clinical situations, may be applied to the diagnosis of skeletal lesions from archeological specimens. Despite the numerous differences between the dry bone specimens obtained from ancient skeletal remains and skeletal lesions in live specimens, the clear use of established radiological terminology to describe basic x-ray patterns is invaluable in enabling paleoradiologists to adopt a more evidence-based approach.

There are few paleopathology textbooks available that deal with a wide spectrum of skeletal diseases. It is not our intention to thoroughly review all skeletal pathologies, but instead we offer the readers a systematic and logical approach to different groups of common pathologies that are found in skeletons from the archeological records. Finally, the question of "cause of death" has been thoroughly studied in the forensic anthropology literature. It is not the purpose of this book. Instead, this chapter seeks to address the medical discussion of paleopathology, which has so often been neglected in the anthropological and osteoarcheological literature.

4.2
The Paleoradiology Method

4.2.1
General Principles
for X-ray Interpretation

The most important first step for paleopathologists, and others who wish to use paleoradiologic techniques, is to identify and collaborate with a skeletal radiologist who is keen and interested in the history of diseases. However, this is very much a two-way street. The paleopathologist or bioarcheologist should make the radiologist aware of taphonomic changes that affect the nature of dry specimens, directing them to key textbooks dealing with that topic. In return, the bioarcheologist should take advantage of the skeletal radiologist's skills and medical background to apply them to dry specimens (Aufderheide and Rodriguez-Martin 1998; Brothwell and Sandison 1967; Ortner 2003). In North America, the qualification of a skeletal radiologist is obtained through a long period of training in musculoskeletal pathology, which includes 4 years of undergraduate studies in sciences, then 4 years in medical school, followed by 5 years of residency in radiology with an additional fellowship year in skeletal imaging.

For nonradiologists, another important key to appropriate interpretation is to read relevant textbooks of skeletal radiology that are recommended to radiology residents and skeletal radiology fellows during their training (Brower and Flemmings 1997; Edeiken et al. 1980; Forrester and Brown 1987; Freiberger et al. 1976; Griffith 1987; Helms 2005; Keats 1988; Resnick and Kransdorf 2005; Schmorl and Junghans 1956). It is important to re-emphasize here that the best x-ray study for detecting and establishing the differential diagnosis of skeletal lesions in dry bone specimens is conventional radiography (i.e., plain x-ray images). In most cases, CT is only needed to clarify changes already seen on x-ray by suppressing the superimposition of bony structures in complex anatomical areas such the spine, base of skull or pelvis, or soil matrix that may hide some part of the lesion. Finally, it is essential to establish clear communication between radiologist and bioarcheologist in order to clarify and determine the terminology used in radiological reports.

The paleoradiological method is utilized toward the establishment of an accurate diagnosis of bone lesions, based on the appropriate interpretation of radiological findings on a dry specimen as shown on x-ray studies of acceptable quality (Tables 4.1 and 4.2). It is essential to obtain the best x-ray image of the specimen. The technical requirements have been discussed elsewhere in this book. The next important task for the radiologist is not only to identify the basic x-ray patterns of the bony lesion, but also to distinguish them from normal anatomical variants. For this purpose, radiologists usually consult the Dr. Keats' Atlas of normal anatomical variants, which is widely used (Keats 1988). In addition, the paleoradiologist must be aware of the taphonomic changes that may simulate authentic bone disease (Table 4.3). The confounding factors during the x-ray pattern recognition process include soil matrix, other material deposits on the bone surface, and/or any other postmortem bone and joint alteration (Brothwell and Sandison 1967; Rogers and Waldron 1995; Ruhli et al. 2004; Steinbock 1976a).

All bone lesions can be categorized as destructive, formative, or more commonly, a combination of both (Resnick and Kransdorf 2005). Their shape, size, number, and location within the bone and their distribution within the skeleton are important considerations in the differential diagnosis.

When discussing a differential diagnosis in clinical situations, the number of lesions provides clues to the final diagnosis. Therefore, distinguishing a solitary lesion from multiple lesions is essential and can be achieved easily by interrogating the patient or by doing a nuclear medicine bone scan. It is much more challenging when dealing with dry bones, as the collection does not always include an entire skeletal assemblage. In addition, a few bones from a complete assemblage may be of

Table 4.1. The paleoradiological method of diagnosis

1.	Obtain the best x-ray image of the specimen
2.	Identify the lesion
3.	Analyze systematically the basic x-ray patterns of the lesion
4.	Combine relevant basic x-ray patterns
5.	Determine if the x-ray abnormality is a normal anatomical variant
6.	Determine if the x-ray abnormality is the result of taphonomic alteration
7.	Discuss the differential diagnosis
8.	Always discuss pseudopathology
9.	Suggest the final diagnosis from the broad category of bone and joint diseases

Table 4.2. Lesions within individual bones: basic x-ray patterns

1.	Radiolucent, radiodense, or mixed
2.	Shape
3.	Size
4.	Solitary or multiple
5.	Location within the individual bone
6.	Distribution within the skeleton

Table 4.3. Dry bones: factors affecting x-ray pattern recognition

Bones	Quality of postmortem preservation Taphonomic alteration Soil matrix or other materials on bone surface No soft tissues
Joints	Anatomical alignment Disarticulated
Skeleton	Complete assemblage Incomplete Mixed with bone from other skeletons
Background information	No clinical history available No laboratory tests available

too poor quality to be radiographed, and this limitation must be kept in mind in paleoradiology diagnoses.

Another point of importance is the difference between ancient skeletal remains and mummies' bones. In contrast to ancient skeletal remains where the set of bones may be incomplete, most mummies present with a complete skeleton, covered by desiccated soft tissues. The advantage of having a complete skeleton is, however, offset by the difficulty in reading the x-rays, due to the size of the skeleton, the presence of the wrapping, and the position of the arms, particularly when they are crossed on the chest, which leads to the superimposition of bone over bone. In these cases, CT imaging, with its cross-sectional ability, may play a major role in bony evaluation.

4.2.2
The Classification of Human Bones

The classification of bones (Table 4.4) is essential to understand bone functions and lesions, because parameters such as the shape and structure are due to genetic, metabolic, and mechanical factors. Bones act as levers (involved in trauma), storage for calcium (altered in rickets and osteomalacia), and bone marrow space (disturbed by infection, tumor or anemia). Beyond the pathogenesis, classification provides clear terminology that allows effective communication between experts from diverse backgrounds collaborating in paleopathology studies.

4.3
Gamuts Approach: The Tricks of the Trade

The essential concepts that have been addressed in this chapter include the analysis of basic x-ray patterns, radiopathological correlation, and the target approach to arthritis and the gamut of bone and joint

Table 4.4. Classification of human bones

Bone classification	Examples
Long bones	Femur
	Tibia
	Fibula
	Ulna
	Humerus
	Phalanges
	Metacarpals
Short bones	Tarsals
	Carpals
Flat bones	Ribs
	Sternum
	Scapula
	Skull bones
Irregular bones	Vertebrae
	Facial bones

diseases (Chapman and Nakielny 2003; Resnick and Kransdorf 2005; Rogers and Waldron 1989, 1995; Rogers et al. 1987). Finally, one must understand that the use of clear terminology is essential for efficient communication between paleoradiologists and paleopathologists in order to establish the foundation of an evidence-based approach to paleopathology, without which the study of ancient skeletal diseases, both at the individual and population levels, will be heavily flawed.

Instead of a long and narrative text, tables of differential diagnosis or gamuts are provided to assist the paleopathologist who may have little or no medical background, in understanding the principles behind radiological interpretation of bone diseases as shown on x-rays images. Given the numerous pathologies that affect the skeleton, the list of gamuts provided in this chapter is indeed not exhaustive. Readers who are willing to further explore the capability of diagnostic paleoradiology are invited to consult the numerous basic textbooks listed in the bibliography below.

4.3.1
The Classification of Human Joints

Classification of joints helps to establish terminology based on their anatomical structure, shape, and function. Joints are divided into three main categories. Synarthrosis joints are seen in skull sutures, syndesmosis at the ankle, and gomphosis of the teeth and bone sockets. Amphiarthroses are joints that are connected by ligaments or elastic cartilage like the discovertebral joint. Diarthroses are synovial joints of the limbs and the spine (uncovertebral, facet, and costovertebral joints).

Cross-sectional imaging (i.e., CT) modalities can depict the joints in the axial, coronal, and sagittal planes. Perhaps the easiest ways to visualize these are: axial images divide the anatomy into upper and lower parts, coronal divide the anatomy into front and back parts, and sagittal divide the anatomy into left and right parts. In addition, advanced imaging software can reconstruct high-resolution imaging data into three-dimensional views.

4.4
Bone Trauma

X-ray examination is the first test used for detecting trauma in paleopathology and in a clinical context (Crawford Adams and Hamblen 1999; Galloway 1999; Griffith 1987; Resnick and Kransdorf 2005; Schultz 1991; Steinbock 1976a).

4.4.1
The Classification of Fractures and Basics of X-ray Interpretation

X-ray study is the best first-line test to diagnose and classify fractures (Table 4.5). The first step is to define a fracture as incomplete or complete. A fracture is complete when there is a complete discontinuity of the bone. In a closed or simple fracture, the skin is intact. When the fracture communicates with the outside environment by breaking through the skin, it is called an open or compound fracture. This type of fracture is much more prone to infection and is associ-

ated with severe morbidity and high mortality if left untreated. The next diagnostic step is to describe the site of the fracture. It is important to use anatomical terms for fracture description, such as femoral neck, base of the metatarsal, waist of the scaphoid, ulnar styloid, or medial malleolus. A long bone is divided in proximal, middle, and distal thirds. Following these first two steps, one must describe the fracture lines. In a transverse fracture, the line is perpendicular to the long axis of the cortices. In oblique fractures, the line runs obliquely to the cortex axis. A spiral fracture is an oblique fracture that spans the circumference of the bone. A comminuted fracture has more than two

Table 4.5. Fracture terminology in a clinical context

Main categories	Acute
	Stress
	Pathologic
Anatomic site	For long bones, divide the shaft into thirds
	Standard anatomic terminologgy use (surgical neck of the humerus, lateral condyle)
Pattern of fracture	Simple (two fragments)
	Comminuted (more than two fragments)
Direction of fracture line[a]	Transverse
	Spiral
	Oblique
	Longitudinal
Apposition and alignment[b]	Displacement (medial, lateral, posterior, anterior, dorsal, volar)
	Angulation (medial, lateral, posterior, anterior, dorsal, volar)
	Rotation (internal, external)
	Distraction (separated fragments)
	Avulsion (fragment separated by pull of a ligament/tendon)

[a] Does not apply to paleoradiology if the fracture occurred in the perimortem period because postmortem changes may further displace the two fragments of the fractures
[b] Most patterns do not apply to paleoradiology. Alignment applies only if the fracture heals premortem in a wrong position, called malunion

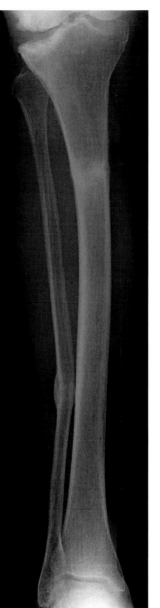

Fig. 4.1. Stress (insufficiency) fracture at the proximal tibia and distal fibula: clinical case

bony fragments. Displacement is described by defining the position of the distal fragment with respect to the proximal one. Angulation is defined in degrees varus or valgus with respect to the midline. Compression fracture is used in spinal trauma. Depressed fracture can be used for the tibial plateau or skull fracture.

Fracture in children should be classified differently, as children are not "small adults," but have a distinctive skeletal trauma pattern. Greenstick fractures are seen in children between 5 and 12 years old. One cortex is broken, while the other is bent. Torus fracture is an impaction fracture seen in young children. It occurs most commonly in the distal radius (Aegerter 1975; Schultz 1991).

Beyond these two main groups, fractures can be divided into three types: (1) acute fracture, (2) stress fracture (Figs. 4.1 and 4.2), including fatigue fracture (abnormal stress on normal bone; e.g., military recruits), and insufficiency fracture (normal stress on abnormal bone; e.g., elderly with osteoporosis), and (3) pathologic fracture, where there is underlying bone pathology.

cheological record: antemortem (before death), perimortem (near the time of death), and postmortem (after death) fractures (Roberts and Manchester 2005) (Table 4.6). Distinguishing antemortem from peri- and postmortem is easy because of evidence of healing (i.e., presence of a callus in antemortem fractures). Differentiating perimortem (no callus formation yet) from postmortem fractures is much more challenging, and sometimes impossible (Lovell 1997). The analysis of the site, pattern, and direction of the fracture line is essential, as is knowledge of fracture mechanisms. In most cases, postmortem fractures have sharp margins and do not fit with any well-identified types described in the classification used in the clinical context. The description of the mechanisms and types of fractures is beyond the scope of this book. Readers are urged to consult the various orthopedic surgery and forensic pathology textbooks listed in the bibliography. A consultation with an orthopedic surgeon, forensic pathologist, and musculoskeletal radiologist is essential to interpret challenging fractures in archeological settings.

4.4.2
Differential Diagnosis

The major problem here is not to differentiate a fracture from any other types of bone lesion, but instead to identify the three types of fractures from the ar-

4.4.3
The Healing Process and Complications of Fractures

The healing process is divided into five successive histologic stages: hematoma, cellular proliferation, callus, consolidation, and remodeling (Table 4.7).

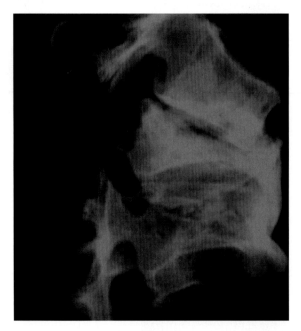

Fig. 4.2. Compression fracture in osteoporosis. Prehistoric Amerindian. Berkeley collection

Table 4.6. Antemortem versus perimortem versus postmortem fracture

1. Antemortem	Sign of healing: callus with good alignment or malunion
2. Perimortem	Fracture line (spiral, greenstick, depressed pattern)
3. Postmortem	Many small fragments, no callus formation, no clear pattern or fracture line

Table 4.7. Tubular bone healing process

1.	Hematoma
2.	Cellular proliferation
3.	Callus: woven bone
4.	Consolidation: lamellar bone
5.	Remodeling

The hematoma is contained by surrounding muscles, fascia, and skin. Subperiosteal and endosteal cellular proliferation starts at the early phase, beginning from the deep surface of the periosteum and progressing toward the fracture. It is during this stage that repair tissue is formed around each fracture fragment, and then gradually progresses to bridge the gap between the two fragments. This immature bone, called osteoid or woven bone, forms a solid bone mass in and around the fracture site, which is called the primary callus. At this stage the callus become visible on x-rays. During the consolidation stage, the primary callus is transformed into lamellar bone under the combined action of osteoblasts and osteoclasts. The remodeling stage is the last step of fracture healing. Bone is strengthened in the lines of stress; the rest is resorbed. A fracture in children heals within 4–6 weeks. In adults the period is longer and may take up to 6 months. Radiological criteria for evidence of fracture healing are a callus that bridges the fracture site with continuity of trabecular bone across the fracture (Figs. 4.3–4.5).

Healing of cancellous bone is different form that of tubular bone. The repair process occurs between the two bone surfaces without the need of callus formation.

There are two main types of complications of fractures: fracture-related and injury-related. Fracture-related complications include infection, delayed union, malunion, nonunion, or avascular necrosis. Injury-related complications include life-threatening injuries to major vessels, injury to the viscera and/or neurovascular structures, or fat embolism. All of these complications were lethal before the era of modern medicine.

Depending on their etiology, site, and concomitant injuries, fractures carry a variable degree of morbidity (Table 4.8) that may lead to death without proper

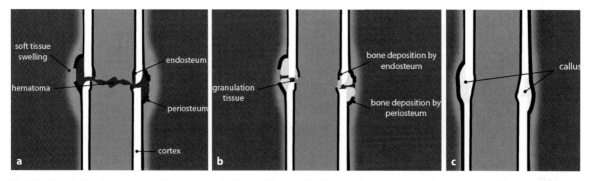

Fig. 4.3 a, b, c. Callus formation. **a** Hematoma, cellular proliferation (neither the hematoma nor the cellular proliferation is seen on x-rays). **b** Consolidation. The callus is seen on x-rays, but granulation tissues are not. **c** Callus and bone remodeling

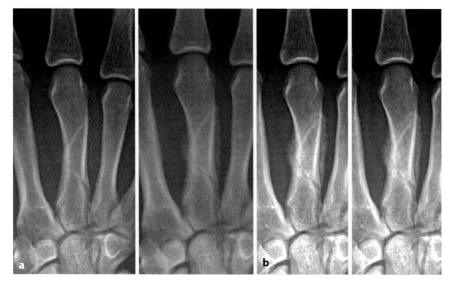

Fig. 4.4 a, b. Fracture of the third metacarpal. Sequence of a healing process. **a** Day 1; **b** day 14; **c** day 28; **d** day 42

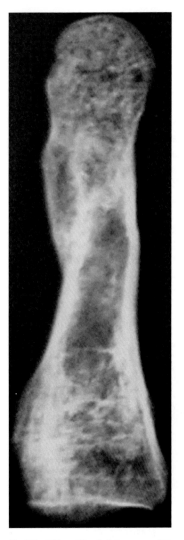

Fig 4.5. Old and healed fracture of the metacarpal

Table 4.8. Complications of fracture

Immediate	Sign of healing: callus with good alignment or malunion
	Acute ischemia[a]
	Hemorrhagic shock[a]
	Fat embolism[a]
	Injury to local structures: nerves, muscles, tendons[a]
Delayed	Infection[b]
	Delayed union[b]
	Nonunion[b]
	Malunion[b]
	Avascular necrosis[b]
	Limb shortening[b]
	Ischemic contracture[c]
	Regional/local osteoporosis[c]

[a] Not seen in paleoradiology. Only fractures that have had no time to heal will be seen
[b] Seen in paleoradiology
[c] Not obvious

modern treatment. In ancient times, vascular complications of fractures such as acute ischemia, hemorrhagic shock, or fat embolism were fatal. However, without proper treatment, some fractures may heal, while others may result in complications such as nonunion, delayed union, malunion, limb shortening, or avascular necrosis, which allow the patient to survive but with permanent morbidity (Figs. 4.6–4.9). Infection is an important cause of morbidity and will be discussed later in this chapter (Milgram 1990; Schultz 1991). Other trauma to bone includes foreign body (Fig. 4.10), trepanation (Fig. 4.11), heterotopic bone formation (the term myositis ossificans is a misnomer; Fig. 4.12), and amputation (Fig. 4.13).

4.5
Joint Trauma

Traumatic joint injuries can be divided into three types: dislocation, subluxation, and diastasis. A dislocation is a complete disruption of the joint with loss of contact between the articular cartilages. A subluxation is a partial loss of contact of opposing articular surfaces. Diastasis is a traumatic separation of a fibrocartilage joint, such as the symphysis pubis (Aegerter 1975; Griffith 1987; Resnick and Kransdorf 2005; Schultz 1991; Steinbock 1976a). This clinical definition of joint injuries may not be applied perfectly in paleopathology, as joints may be altered significantly by taphonomic changes when one tries to assess a complete skeletal assemblage in situ in the soil matrix. Once the bones are removed, the joints "disappear" as the bones are collected separately. However, old and neglected joint dislocation may alter the structures of opposing articular surfaces that may be detected by both observation and x-ray study.

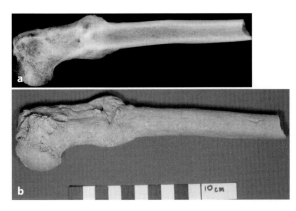

Fig. 4.7 a, b. Healed fracture of the proximal humerus with exuberant callus

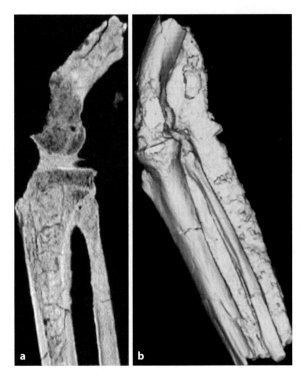

Fig. 4.8 a, b. a Supracondylar fracture of the femur/malunion: two-dimensional (2D) computed tomography (CT). **b** Supracondylar fracture of the femur/malunion: three-dimensional (3D) CT. a and b: specimen from 2000-year-old Prei Khmeng, Angkor, Cambodia

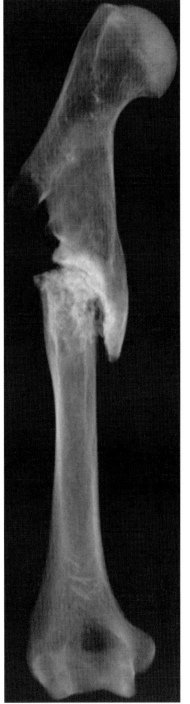

Fig. 4.9. Fracture of the midshaft of the humerus: nonunion with pseudoarthrosis. Prehistoric Amerindian Berkeley collection

4.6
Arthropathies

4.6.1
Classification of Arthropathies

A fundamental understanding of the classification of arthropathies is necessary for paleopathologists, who struggle to bring sound methodology in order to develop an "evidence-based" discipline, which is essential in supporting the trend toward a population study of ancient diseases (Roberts and Manchester

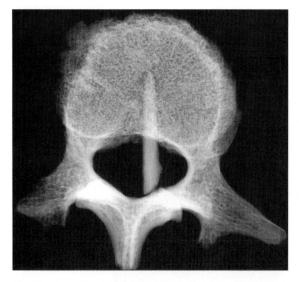

Fig. 4.10. Arrowhead through the spinal canal and vertebral body. Prehistoric Amerindian, Illinois (courtesy of the late Dr. Morse)

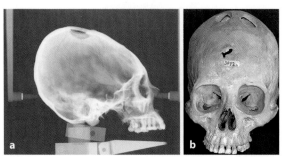

Fig. 4.11 a, b. Skull with trepanation (courtesy of Professor Villari)

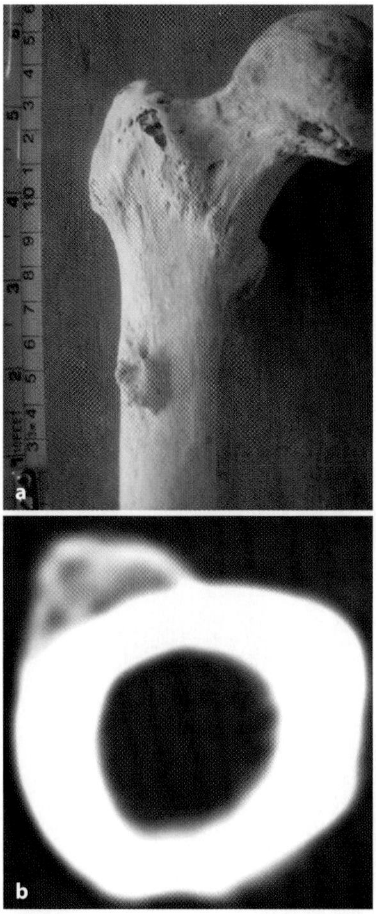

Fig. 4.12 a, b. Heterotopic bone formation (myositis ossificans). Samrong Sen, prehistoric Cambodia

2005; Rogers et al. 1987, Rogers and Waldron 1995; Steinbock 1976a). The major problem in the current approach to bone and joint diseases is the lack of a clear and standard terminology agreement between experts from diverse backgrounds. To address this crucial issue, we propose here a classification based on broad categories of arthropathies used in clinical situations (Brower and Flemmings 1997; Edeiken et al. 1980; Freiberger et al. 1976; Resnick and Kransdorf 2005) (Table 4.9). Similar to the diagnostic approach suggested by Rogers et al. in the interpretation of dry joint specimens, skeletal radiologists first identify the basic patterns of x-ray changes in joint disorders, and then assess the distribution of those basic abnormalities within a joint, and then the entire skeleton (Rogers and Waldron 1995). For instance, basic abnormalities of the knee will be described at each knee compartment (i.e., the femoropatellar and the lateral and medial femorotibial compartments).

Following this step, the other knee will be evaluated to determine if there is symmetric or asymmetric involvement. Finally, other joints of the entire skeleton in the extremities, spine, and pelvis are checked for any abnormalities. Like many other bone and joint disorders, x-ray study is the most accurate first-line test to be performed. Also, as suggested by Rogers and Waldron (1995), "….the advice of an experienced skeletal radiologist is absolutely crucial in drawing conclusions from the films which are taken." As erosion is a major radiological criterion in the differential diagnosis of arthropathies, its definition should be stated clearly upfront. "Erosive arthropathy" is a term widely used in the paleopathology literature, but it should be applied with caution as in clinical, pathological, and radiological paradigms, this may mean an arthropathy with erosions, which is not specific, in contrast to "erosive osteoarthritis," which is a well-defined clinical entity.

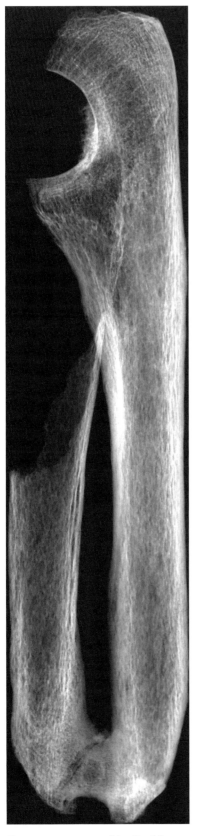

Fig 4.13. Amputation of the distal forearm

Table 4.9. A classification of bone and joint diseases based on broad categories of arthropathies, used in clinical situations

Type	Subtype	Examples
Degenerative	Primary	Age-related
	Secondary	Post-trauma
Inflammatory	Seropositive	Rheumatic arthritis, scleroderma, systemic lupus erythematosus, dermatomyositis
	Seronegative	Ankylosing spondylitis, Reiter's syndrome, psoriasis, inflammatory bowel disease
Metabolic	Crystal deposition	Gout, CPPD, hydroxyapatite
	Other depositional	Hemochromatosis, Wilson's disease, hemophilia
	Endocrine	Acromegaly
Infectious	Bacterial, viral, fungal	Osteomyelitis, septic arthritis, Tuberculosis
Neuropathic (Charcot's joint)		Diabetes, tabes dorsalis, spinal cord injury

It is important to reiterate that effective and clear communication is essential between experts dealing with paleopathology. Another term that creates much confusion in the literature is "exostosis" to define an ossified enthesopathy. The latter term is much more appropriate when dealing with arthropathies, while the former term actually stands for a benign tumor, osteochondroma. Also, "periostitis" means inflammation of the periosteum. It is therefore appropriate to use this term in bone infection or periostitis associated with seronegative arthropathies such as ankylosing spondylitis, psoriatic arthritis, Reiter's disease, or arthritis associated with inflammatory bowel disease. However, x-ray films cannot determine inflammation of the periosteum, as they are not specific. To prevent any confusion, it is wiser to use the term "periosteal reactions," which is more appropriate in paleoradiology. It also can be applied to other bone disorders that are not classified as inflammation or infection such as, for example, callus in fracture, eosinophilic granuloma, and Ewing tumor. Finally, the many "phytes" of the spine must be defined appropriately because they alter the differential diagnosis of spine disorders. This issue will be discussed later in this chapter.

4.6.2
Basics of X-ray Interpretation

In clinical situations, x-ray interpretation follows the ABCs: alignment, bony mineralization, cartilage space and soft tissues (Forrester and Brown 1987). This systematic approach allows for an accurate diagnosis of arthropathies based on the assessment of plain x-ray films (Brower and Flemmings 1997; Forrester and Brown 1987; Resnick and Kransdorf 2005) (Table 4.10). Unfortunately, the ABCs are not entirely appropriate when applied to paleoradiology, as dry skeletons from archeological records have lost their soft tissues and joint alignment, and so the first and last parameters are inapplicable. Also, because taphonomic processes may have altered the chemical composition of the bones, extreme caution should be used when analyzing bone mineralization, and cartilage space assessment is also limited for these same reasons. To further confuse the diagnosis, soil matrix may have collected within the joint space, leading to a widening of that space. In summary, the logical approach to arthropathies widely used in clinical radiology has some limitations when applied to paleopathology, which should be taken into consideration when one is involved in paleoradiological diagnosis.

What then to do with the dry joint? An alternative method is to base the x-ray interpretation on the so-called "target area approach," which assesses the distribution of basic x-ray abnormalities within each joint of the skeleton. This approach, first proposed by Resnick, allows quite an accurate diagnosis of arthropathies when combined with the analysis of basic pattern changes (Resnick and Kransdorf 2005) (Figs. 4.14–4.23). However, this approach is possible only if the entire skeletal assemblage is complete, which is not always the case in paleopathology. When a single joint is abnormal, then the gamut of diagnostic considerations would fall under the category of monoarthropathies (infection, gout, or posttraumatic osteoarthritis). If there are multiple joints involved, then the approach includes differential diagnoses in the polyarthropathy category. In this situation, all joints must be assessed (i.e., the peripheral joints and those of the spine). For this approach to be valid, it is essential to be able to identify and recognize the basic x-rays patterns seen in inflammatory, degenerative, and metabolic arthropathies (Tables 4.11–4.14). Indeed, there is always the possibility of overlap between these x-ray patterns.

Table 4.10. Basic x-ray patterns in joint diseases

Bone formation	Osteophytes (margin of articular cartilage)
	Syndesmophytes (annulus fibrosis/spine)
	Enthesophytes (ligament and fascia)
	Subchondral bone sclerosis
	Periosteal reactions
Bone destruction	Subchondral cyst
	Articular erosions (central/marginal)
	Periarticular erosions
Joint space narrowing[a]	
Malalignment[a]	
Periarticular soft-tissue swelling[a]	

[a] Not applicable to paleoradiology

Table 4.11. Degenerative joint disease

1.	Subchondral cyst
2.	Subchondral sclerosis
3.	Osteophytes
4.	Joint space narrowing[a]

[a] Not applicable to paleoradiology

Table 4.12. Inflammatory arthritis (limbs)

1.	Articular erosions
2.	Bony ankylosis
3.	Periosteal reactions
4.	Enthesopathy
5.	Periarticular osteoporosis[a]
6.	Joint space narrowing[a]
7.	Subluxation[a]

[a] Not applicable to paleoradiology

Table 4.13. Inflammatory arthritis (spine/pelvis)

1.	Syndesmophytes
2.	Squaring of vertebral body
3.	Sacroiliitis

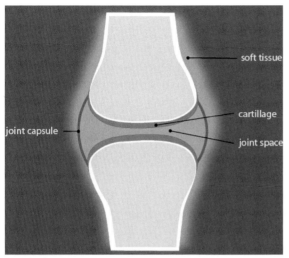

Fig 4.14. Normal synovial joint

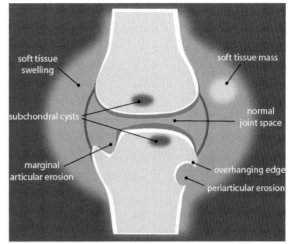

Fig. 4.17. Crystal-induced arthropathy (gout and others)

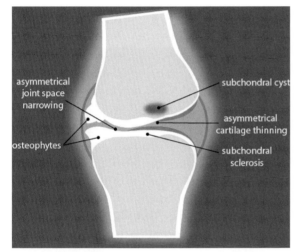

Fig. 4.15. Osteoarthritis

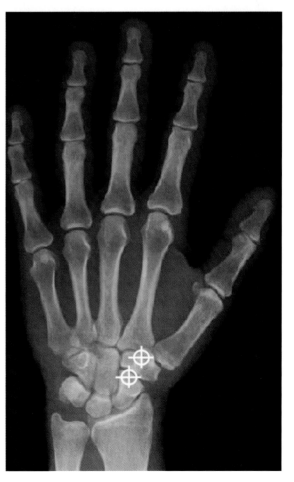

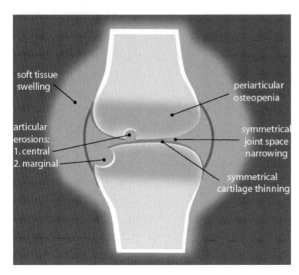

Fig. 4.16. Rheumatoid arthritis

Fig. 4.18. Target approach. Wrist: osteoarthritis

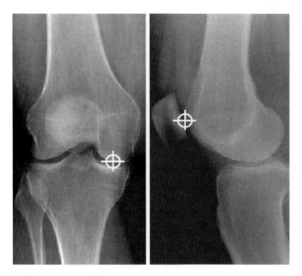

Fig. 4.19. Target approach. Knee: osteoarthritis

Table 4.14. Bony overgrowth of the spine: spinal arthropathies

Syndesmophytes	Ankylosing spondylitis
	Psoriatic arthropathy
	Reiter's disease
	Inflammatory bowel diseases
Enthesophytes	Diffuse idiopathic skeletal hyperostosis
Osteophytes	Spondylosis deformans
	Degenerative disc disease

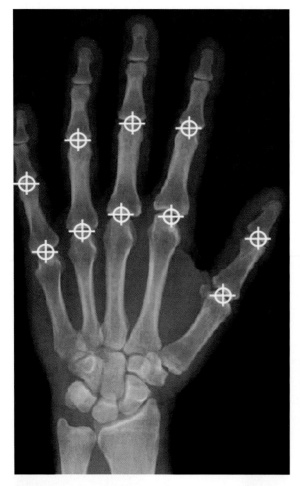

Fig. 4.20. Target approach. Hand: rheumatoid arthritis

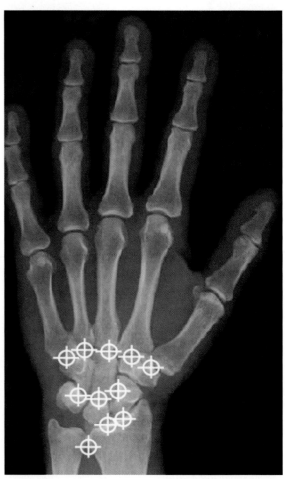

Fig. 4.21. Target approach. Wrist: rheumatoid arthritis, ankylosing spondylitis, psoriatic arthritis, Reiter's syndrome, and gouty arthritis

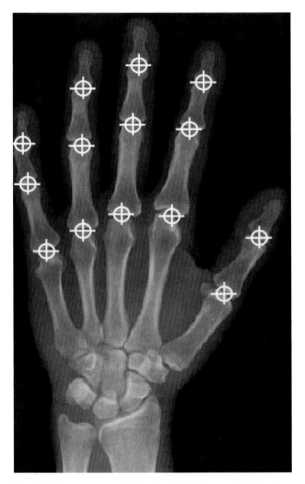

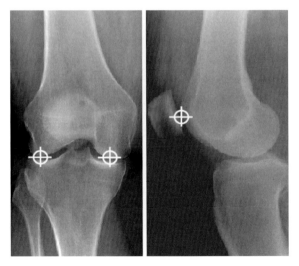

Fig. 4.22. Target approach. Hand: ankylosing spondylitis, psoriatic arthritis, Reiter's syndrome, osteoarthritis, and Gouty arthritis

Fig. 4.23. Target approach. Knee: rheumatoid arthritis, ankylosing spondylitis, psoriatic arthritis, Reiter's syndrome, and gouty arthritis

4.6.3
Arthropathies of the Spine and Pelvis

There are two main types of degenerative process in the spine (Resnick and Kransdorf 2005; Schmorl and Junghans 1956), which can cause acute and chronic back pain. Spondylosis deformans is a degenerative change of the "annulus fibrosus" of the intervertebral disc with osteophyte formation at the anterior margin of the vertebral body. It is defined as a bony outgrowth from the point of attachment of the annulus fibrosis at the margin of the vertebral body, which initially develops horizontally but then turns vertically. Osteophyte formation is the result of Sharpey fiber tears with disruption of the normal connection between the disc and the vertebral body. The major x-ray sign of spondylosis deformans is a spinal osteophyte, with preservation of the disc space. This osteophyte is much less exuberant than the bony overgrowth seen in diffuse idiopathic skeletal hyperostosis (DISH) called enthesophytes, which is described later in this section (Fig. 4.24).

Degenerative disc disease involves the central portion of the disc called the nucleus pulposus. X-ray findings include disc-space narrowing, osteophytes, and sometimes a Schmorl's node, which is an intraosseous displacement of discal material through a defect of the cartilaginous endplate and subchondral plate of the vertebral body. A Schmorl's node is caused by repetitive trauma, which produces a herniation of the nucleus pulposus through a weak point of the endplate. This weak spot is the result of a combination of three factors:

1. The cartilaginous endplate is thinner at the passage of the former notochord that has resorbed.
2. Presence of a few residual vascular canals of the discal vessels that have resorbed.
3. Presence of minute multiple spots of bone necrosis in the area (Schmorl and Junghans 1956).

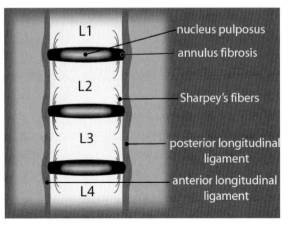

Fig. 4.24. Normal anatomy of the lumbar spine: lateral view

DISH is defined as an ossification of entheses that includes ligament and tendon attachment over the bone. The most common site of enthesopathy is the spine, but it can also affect any other entheses of the skeleton. Unlike ankylosing spondylitis, where it is called syndesmophyte, there is no ankylosis (stiffening or fusion of a joint) of the synovial part of the sacroiliac joints or facet joints, and DISH is most often asymptomatic (Table 4.15; Fig. 4.25). Histologic study shows ossification of the anterior longitudinal ligament, paravertebral connective tissues and the annulus fibrosus. In a thorough review of the literature on paleopathology, Rogers et al. showed that many specimens previously reported as ankylosing spondylitis are actually examples of DISH, as this latter spine disorder is commonly interpreted as ankylosing spondylitis (Rogers et al. 1987; Rogers and Waldron 1995).

The most striking example of this medical error in mummy studies was the case of Ramsesses II (Chhem et al. 2004) (Fig. 4.26–4.29).

Seronegative spondyloarthropathies are mainly a disease of the enthesis of the pelvis, spine and extremities (Fig. 4.30–4.32). This broad group includes ankylosing spondylitis, psoriatic arthritis, Reiter's disease, and arthritis associated with inflammatory bowel disease. They involve primarily the axial skeleton, namely the spine and the sacroiliac joints, but they can also affect the extremities. In summary, a seronegative spondylarthropathy is a group of polyarthropathies made of a combination of sacroiliitis, spondylitis, and peripheral arthritis. Although there

Table 4.15. Diffuse idiopathic skeletal hyperostosis – Resnick's criteria

1.	Ossification of the anterior longitudinal ligament over four contiguous vertebral bodies
2.	No facet joint fusion, no sacroiliitis, no sacroiliac fusion
3.	Disc space is preserved[a]

[a]Not applicable in paleoradiology

Table 4.16. X-ray patterns of seronegative spondylarthropathies

Sacroiliac	Articular erosions (iliac side first then sacral side)
	Subchondral bone sclerosis
	Ankylosis
Spine	Squaring: erosion of the superior and inferior corners of the anterior margin of the vertebral body, which becomes "straight"
	Shiny corner: increased density of the anterior corners of the vertebral body due to osteitis
	Syndesmophytes
	Spinal fusion
Extremities	Erosion: central or marginal
	Bone proliferations: – along the shaft: periosteal reactions – across the joint: ankylosis – along the entheses: enthesopathy – at the edge of marginal erosions
	Preservation of bone density[a]

[a]Difficult to confirm in dry specimens

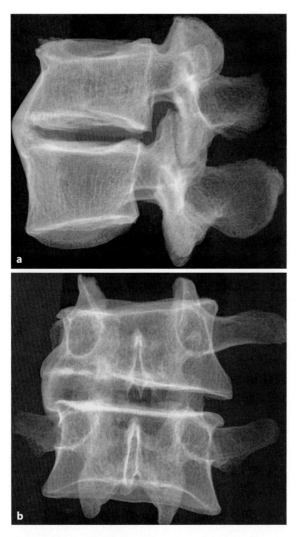

Fig. 4.25 a, b. Enthesophyte, diffuse idiopathic skeletal hyperostosis (DISH; courtesy of Dr. El Molto)

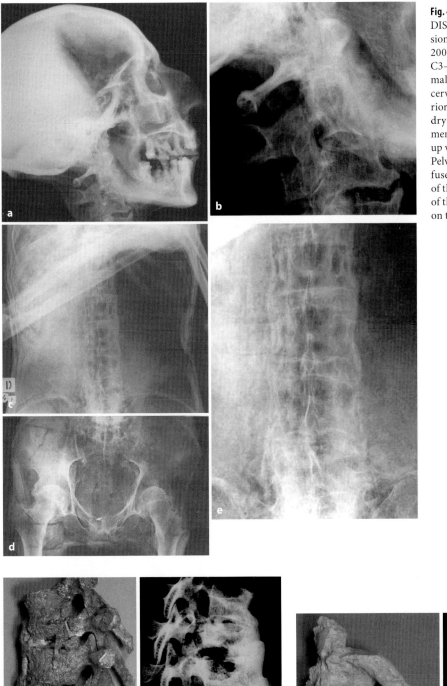

Fig. 4.26 a–e. Ramesses II. DISH (reprinted with permission from CAR Journal, June 2004). **a** Enthesophyte at C3–C4. Facet joints are normal. **b** Close-up x-ray of upper cervical spine. **c** Anteroposterior view of abdomen. Normal dry anterior longitudinal ligament of lumbar spine. **d** Close up view of lumbar spine. **e** Pelvis: sacroiliac joints are not fused. Ossified enthesopathy of the insertion of the tendons of the rectus femoris (best seen on the right side)

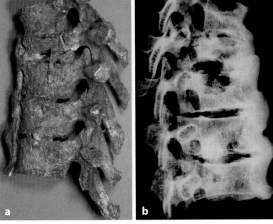

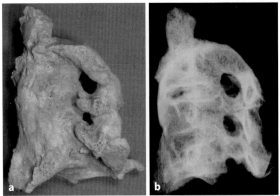

Fig. 4.27 a, b. DISH at upper thoracic spine in dry specimen. Thick enthesophytes. Facet joints are not fused

Fig. 4.28 a, b. DISH at cervical spine in dry specimen

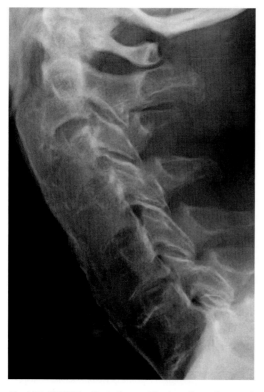

Fig. 4.29. DISH in clinical case. Thick enthesophyte – the facet joints are not fused

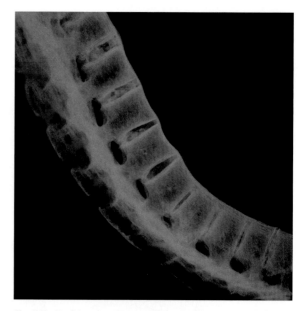

Fig. 4.31. Prehistoric Amerindian. Crable site. Ankylosing spondylitis in a dry spine specimen. Note the thin syndesmophyte along the anterior aspect of the spine and fusion of the facet joints (courtesy of the late Dr. Morse)

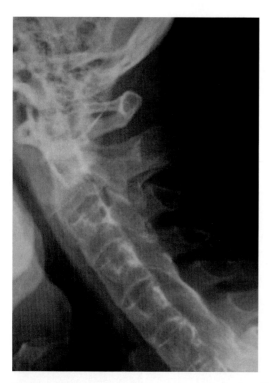

Fig. 4.30. Ankylosing spondylitis of the cervical spine. Clinical case. Facet joints are fused. Thin syndesmophyte along the anterior aspect of the spine

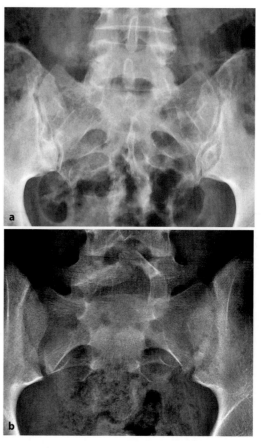

Fig. 4.32 a, b. a Normal sacroiliac joints. **b** Sacroiliitis of left sacroiliac joint in ankylosing spondylitis (clinical case)

are no pathognomonic x-ray patterns, the combination of clinical and radiological patterns may help in the differential diagnosis of each of these four spondylarthropathies (Brower and Flemmings 1997; Resnick and Kransdorf 2005) (Table 4.16). Psoriatic arthritis is associated with skin disease (psoriasis). Reiter's disease can be sexually transmitted or is associated with a postdysentery, infectious "enteropathic" arthropathy with gastrointestinal and liver diseases (Crohn's disease, ulcerative colitis, Whipple's disease, or biliary cirrhosis). Again, the prevalence of spondylarthropathy is lower in the archeological records than it was previously described because of the confusion with DISH (Chhem et al. 2004; Rogers and Waldron 1995).

4.6.4
Arthropathies Affecting the Limbs

Rheumatoid arthritis (Brower and Flemmings 1997) is a disease of the synovium that typically affects the hand and wrist joints. This distribution represents a major challenge as many small joints of the hands may have disappeared due to postmortem changes (Rogers and Waldron 1995; Rogers et al. 1987).

Articular erosions are common, with a symmetrical distribution. The earliest site of erosion is at the "bare area," which is the segment of the epiphysis that is not covered with articular cartilage, but is still located within the joint capsule. Early erosions are very subtle at the beginning of the disease and should be searched for carefully, and it is here that posterior-anterior, lateral, and oblique views of the hand and wrist will be most useful. The three other cardinal signs of rheumatoid arthritis that are observed in clinical cases are joint-space narrowing, soft-tissue swelling, and periarticular osteoporosis. Joint-space narrowing and soft-tissue swelling are, of course, lacking in dry joint specimens, and periarticular osteoporosis may be extremely difficult to confirm in dry bones, since density depends on the interaction with the chemical nature of the soil matrix. Alignment of joints in the context of paleopathology is extremely difficult to evaluate as taphonomic changes have altered the joint anatomy. Therefore, the erosions and their distribution remain the best criteria for the establishment of the diagnosis of rheumatoid arthritis. However, they must be differentiated from the pseudoerosion caused by taphonomic processes of diverse origin. The other sites of involvement of rheumatoid arthritis are the feet, hip, and knee (Brower and Flemmings 1997; Resnick and Kransdorf 2005; Rogers and Waldron 1995).

Gouty arthritis is classified as a crystal-induced arthropathy as a result of the deposition of urate crys-

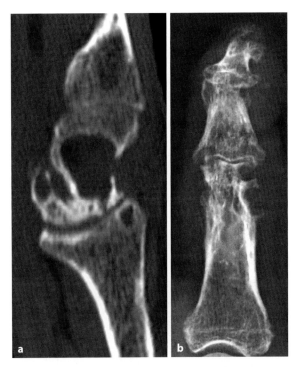

Fig. 4.33 a, b. a Gouty arthritis of the wrist. Radiocapitate joint (clinical case). **b** Gouty arthritis at the interphalangeal joints. Articular and periarticular erosions

tals in the joint. It is the oldest arthropathy described in the literature and was initially called podagra (Brower and Flemmings 1997), which means "attack of the foot." X-ray study is usually normal during the early phase. Radiological findings manifest 6–8 years after the initial attack (Brower and Flemmings 1997; Resnick and Kransdorf 2005; Rogers and Waldron 1995; Rogers et al. 1987). Multiple bone erosions are the radiological hallmark of the disease (Fig. 4.33). They include articular (central and marginal) or periarticular erosions with a sclerotic margin, with an "overhanging edge" in advanced and typical cases. Bone density is preserved, but this criterion is difficult to assess in dry joints from archeological records. The most typical site of involvement is the first metatarsophalangeal joint, followed by the other joints of the feet and hands. Gout tophi, chalky depositions of urates around the joints, typical in clinical cases, have long disappeared in dry bone and joints.

4.6.4.1
Osteoarthritis (Degenerative Joint Disease)

Primary osteoarthritis is a disease of chondrocyte degeneration with no pre-existing joint disease, associated trauma or bone deformity. Secondary osteoarthritis is the result of the aging process, body weight, trauma, and anatomy variation. The clinical history covers years to decades rather than days or months in contrast to inflammatory arthritis. Clinical symptoms include stiffness and swelling that may change with the weather conditions. There is no joint motion restriction in early disease. Stiffness arises at the late stage of osteoarthritis.

Advanced pathological findings include osteophytes, subchondral bone cysts, and eburnation, which indicates bone necrosis. Radiological findings reflect the pathological changes; osteophytes appear as bone formations at the edge of the articular surface. Subchondral cysts are radiolucent, well-defined lesions developing in the subchondral bone. Eburnation is an area of sclerosis of the subchondral bone (Aufderheide and Rodriguez-Martin 2003; Milgram 1990).

Osteoarthritis is a common disease in skeletons from the archeological record. Primary arthritis commonly involves the hip (Fig. 4.34), knees, and base of the thumb joints. Any osteoarthritis occurring at the other joints of the appendicular skeleton should be evaluated for a pre-existing lesion such as trauma or inflammatory arthritis.

Osteoarthritis of the spine involves synovial joints, including the apophyseal, costovertebral, and transversovertebral joints. Radiological findings are the same as those described in the appendicular skeleton, namely osteophytes, subchondral cysts, and sclerosis. Joint-space narrowing is not a reliable finding in dry skeletons because of common taphonomic misalignment of the joints.

4.6.4.2
Rotator Cuff Arthropathy

This is a secondary arthritis of the glenohumeral joint that occurs following a chronic rotator cuff tendonopathy and tear (Fig. 4.36). Radiological signs include osteoarthritis of the glenohumeral joint and osteophytes at the humeral greater tuberosity and acromion. Here again, evaluation of the joint space is unreliable because of postmortem changes at the joint space.

4.6.4.3
Neuropathic Arthropathy

Neuropathic arthropathy is the result of advanced diabetes mellitus, syringomyelia, vitamin B12 deficiency or chronic alcoholism, leprosy, spina bifida, tertiary syphilis, or trauma to the spinal cord. Clinical symptoms are extremely variable, but loss of sensory function around the affected joint is the main common factor. Pathologic findings include multiple periarticular fractures of different ages leading to

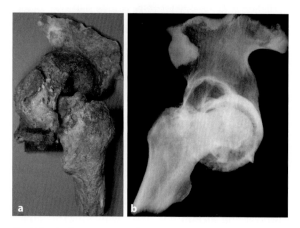

Fig. 4.34 a, b. Severe osteoarthritis, most likely secondary to an unknown arthropathy – with protrusio acetabuli

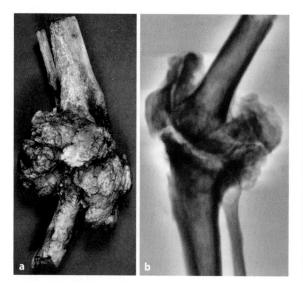

Fig. 4.35 a, b. Neuropathic joint of the knee (courtesy of Professor Vacher-Lavenu)

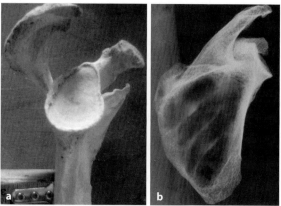

Fig. 4.36 a, b. Rotator cuff arthropathy in a 500-year-old scapula from the Cardamom Mountains, Cambodia

fragmentation of the bones and misalignment of the joints. X-ray findings include fragmentation of bones at the joint, with many loose bodies within the joint space, and misalignment (Fig. 4.35). Multiple joints may be involved in advanced stages of the condition. Sclerosis of the surface of the bone, which borders the joint, may occur (Milgram 1990).

4.7
Infection

4.7.1
Osteomyelitis in Clinical Settings

Bone infections are called osteomyelitis. The acute and hematogenous form occurs more commonly in children. Clinical symptoms include fever, local pain and warmth, restriction of joint motion, irritability, and/or failure to feed. Older children and adults may complain of pain and local swelling, but fever may be absent. When infection affects the spine, back pain is the most common symptom. Before the era of antibiotics, severe forms of osteomyelitis developed into septicemia, which often led to death.

Acute osteomyelitis occurs commonly in children. The clinical findings include pain, restriction of joint motion, fixed posture, swelling, erythema, local warmth, fluctuance of soft tissues, and sometimes pus drainage through a fistula. Irritability and failure to feed are common. Fever may not be present. Before the era of antibiotics, acute osteomyelitis often became chronic, with an indolent course and episodes of fever and draining sinus tracts. Cancer could then arise in those chronic sinus tracts.

In severe and active chronic osteomyelitis, pus may drain continuously through a fistula for many years and decades, and sequestra may be discharged periodically through the skin. Long asymptomatic periods may be followed by recurrent pain, swelling, and drainage. Heel ulcers and sacral decubitus ulcers are common sites of chronic osteomyelitis. Typical chronic osteomyelitis is caused by *Staphylococcus aureus*. However, advanced leprosy or syphilis may also lead to chronic bone infection with swelling, and skin ulcers with pus drainage (Milgram 1990; Resnick and Kransdorf 2005).

4.7.2
Basics of X-rays Interpretation

The x-ray patterns depend on multiple factors including the age of the patient, the virulence of the pathogens, and the stage and severity of the infection. However, the basic x-ray patterns have two main forms:

Table 4.17. Osteomyelitis

Acute osteomyelitis	Lysis
	Periosteal reactions
Chronic osteomyelitis	Lysis
	Sclerosis
	Periosteal reaction[a] (periostitis[b])
	Involucrum[c] (periosteal reactions: live bone)
	Sequestrum[c] (dead bone)
	Cloaca: defect in the cortex (more common in adults)

[a] Nonspecific radiological term
[b] Histopathological term
[c] More common in children

Table 4.18. Spondylodiscitis

1.	Lysis
2.	Sclerosis
3.	Spinal alignment
4.	Disc space (not reliable in dry specimens)

bone destruction and bone formation(Tables 4.17 and 4.18). There are a few radiological findings that are highly specific for chronic osteomyelitis: the presence of a sequestrum, involucrum, and/or cloaca. The presence of these x-ray findings correlate well with the gross pathological findings of chronic bone infection.

4.7.3
Differential Diagnosis

Periosteal reactions and areas of lysis or sclerosis are radiological patterns that are not specific to infection. The differential diagnoses include tumor or pseudotumors (Chapman and Nakielny 2003; Helms 2005; Resnick and Kransdorf 2005). When the lesion is solitary, and especially when it is lytic, the diagnostic consideration should be that of bone tumor until proven otherwise. Bone tumors are discussed elsewhere in this chapter.

4.7.4
Common Bone Infections in the Archeological Record

This section deals with a few types of bone infections selected mostly for their significance in shedding the light on the history of diseases and migration of past populations, at the exclusion of their real prevalence. The most common bone infections described in paleopathology literature are osteomyelitis caused by *S. aureus*, syphilis, tuberculosis, leprosy, and brucellosis (Roberts and Manchester 2005; Rogers and Waldron 1989; Steinbock 1976a).

4.7.4.1
Pyogenic Infection

In the preantibiotic era acute osteomyelitis could not be cured, and evolved into a chronic phase, if the patient survived. Therefore, the presence of a chronic osteomyelitis in the archeological record may testify to the survival of the fittest (Roberts and Manchester 2005). Pyogenic osteomyelitis was the result of poor hygiene, malnutrition, or change in diet during shift from hunter-gatherer status to sedentary agriculture or during the industrial revolution in Great Britain (Steinbock 1976a). It is most commonly caused by *S. aureus*.

Radiological patterns of chronic pyogenic osteomyelitis include osteolysis, osteosclerosis, thick periosteal reaction (involucrum), defect in the cortex (cloaca), and sequestrum, which is a dead bone (Resnick and Kransdorf 2005) (Fig. 4.37).

The establishment of the nature of the infective agent cannot be determined by using x-ray patterns. In clinical situations, the identification of the germ is established by microbiological study of blood samples, pus or materials from bone biopsy. This study of ancient DNA has been proven useful in the diagnosis of causal pathogens of osteomyelitis in skeletal material from the archeological record (Greenblatt and Spigelman 2003). The diagnosis of a typical osteomyelitis is challenging. Differential diagnosis should include bone tumors such as osteosarcoma or Ewing's sarcoma.

Brodie's abscess is a single or multiple radiolucent pus collection within the bone seen during subacute or chronic osteomyelitis. They affect trabecular bones and may occur in the metaphysis, or rarely the epiphysis, of the bone. Sclerosing osteitis of Garré is a chronic infection that affects the cortical bone. Radiologically, osteitis of Garré and Brodie's abscess should be differentiated from osteoid osteoma, a benign bone tumor (Fig. 4.38).

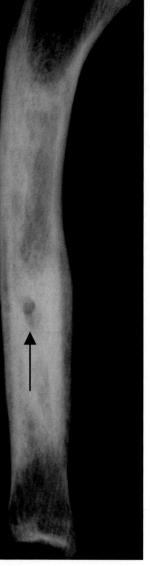

Fig. 4.37. Chronic osteomyelitis. Note a small cloaca at the junction between the middle third and distal third of the tibia. Prehistoric Amerindian from the Berkeley collection

4.7.4.2
Syphilis and other Treponematosis

Syphilis and the other treponematoses are a popular subject of intense debate in the paleopathology literature, where several theories have been proposed to explain the origin and evolution of the disease (Ortner 2003; Powel and Cook 2005; Roberts and Manchester 2005; Steinbock 1976b). The natural history of treponematosis will not be discussed in this section. The focus will be on the radiological diagnosis of syphilis, which is often neglected in paleopathological studies.

Radiological imaging represents a first-line investigative test in both congenital and acquired syphilis (Milgram 1990; Resnick and Kransdorf 2005) (Tables 4.19 and 4.20). Syphilis involves the bone in the later stages of disease (i.e., stage 3). Inguinal adenopathies and chancre appear at stage 1. Hematogenous spread occurs at stage 2. Syphilis is caused by spiro-chetes, namely *Treponema pallidum*, which is sexually transmitted. Spirochetes cause vasculitis, which triggers intense new bone formation, which appears as areas of sclerosis or periosteal reaction, and also some areas of bone necrosis associated with granulation tissue formation called gummata (Aegerter 1975; Milgram 1990; Steinbock 1976a). Bone formation can involve solitary or multiple sites in the skeleton, but is rarely widespread. Skeletal syphilis occurs commonly at the cranial vault (Fig. 4.39), nasal bones, maxilla, and mandible as well as the long bones (tibia, clavicle, hands and feet; Fig. 4.40 and 4.41). X-ray patterns are not pathognomonic and include sclerotic and lytic lesions, periosteal reactions (saber shin at the tibia is classical; Fig. 4.41) and dactylitis at the bones of the fingers. Round, radiolucent areas are foci of active infection or gumma, which are a form of caseous necrosis distinct from that of tuberculosis. It is impossible to radiologically distinguish syphilis from yaws. In keeping with the underlying pathologic process, acquired syphilis appears as periotitis (inflammation/infection of the periosteum), osteitis (inflammation/infection of the cortex), and osteomyelitis (infection of the bone and bone marrow; Resnick and Kransdorf 2005). Their radiological patterns reflect the underlying histopathological changes. Periostitis would be displayed radiographically as periosteal reactions, osteitis as thickening of cortical bone, and osteomyelitis as a combination of the x-ray findings mentioned above associated with multiple radiolucencies in the

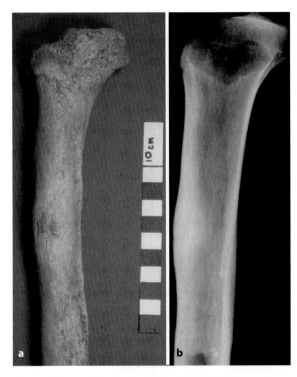

Fig. 4.38 a, b. Focal thickening of the tibial shin. This x-ray pattern is not specific. Differential diagnosis includes chronic osteomyelitis, stress fracture, or osteoid osteoma

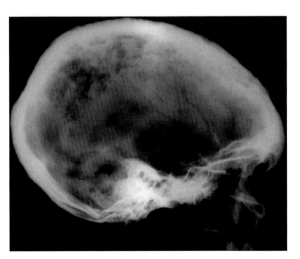

Fig. 4.39. Treponematosis in a postmedieval skull (London, British Museum of Natural History Collection)

Table 4.19. Congenital syphilis

Osteochondritis	Metaphyseal transverse radiolucent band or erosion
	Metaphyseal alternating longitudinal radiolucent and sclerotic band ("celery stalk")
Osteomyelitis	Radiolucent and sclerotic areas, periosteal reactions
Periostitis	Periosteal reaction

Table 4.20. Acquired syphilis: radiological patterns

Osteitis[a]	Cortical thickening[b], increased density[b]
Osteomyelitis[a]	Radiolucent[b] and sclerotic[b] areas, periosteal reactions[b]
Periostitis[a]	Periosteal reactions[b]

[a] Pathological terms
[b] Radiological terms

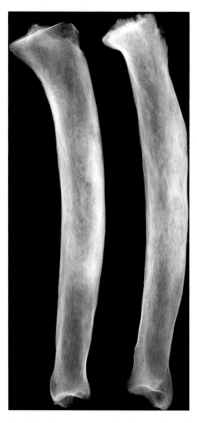

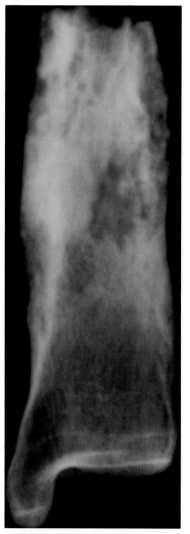

Fig. 4.41. Treponematosis of the distal tibia. Prehistoric Amerindian from the Berkeley collection

Fig. 4.40. Treponematosis of the tibia. "Saber shin" pattern. Prehistoric Amerindian from the Berkeley collection

cortex and trabecular bones. Congenital syphilis in the neonate and the young infant results in osteochondritis, diaphyseal osteomyelitis and periostitis (Resnick and Kransdorf 2005). It is important to stress here that even in clinical situations, x-ray studies can readily detect bony changes caused by treponematosis, but it is not able to determine the causative agent of the infection, which must be diagnosed by a laboratory test. Indeed there are some reports in the literature that support the role of DNA testing in the identification of the specific infectious etiologies resulting in bone infection. DNA testing destroys the bone and so is not widespread as yet; more study is needed to validate the feasibility and accuracy of this test (Herman and Hummel 2003).

4.7.4.3
Tuberculosis

There are four main pathological forms of tuberculosis of the skeletal system: spondylitis, spondylodiscitis, osteomyelitis, arthritis and dactylitis (Resnick and Kransdorf 2005) (Table 4.21; Figs. 4.42–4.46).

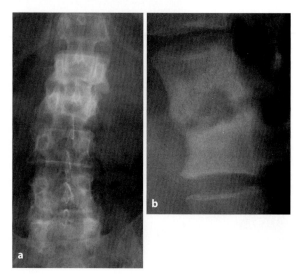

Fig. 4.42 a, b. Spondylodiscitis at L1–L2 (courtesy of Dr. Haddad)

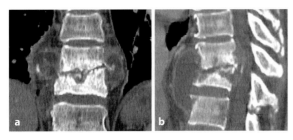

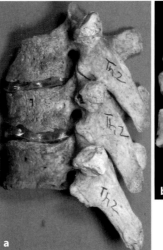

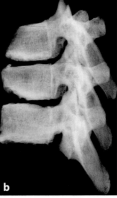

Fig. 4.43 a, b. Tuberculous spondylodiscitis at T10–T11 with parasinal abscess (CT, clinical case; courtesy Professor Wang)

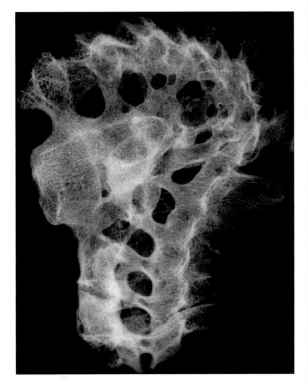

Fig. 4.44. Tuberculosis of the spine. Byzantyne period, Saracane, Turkey

Fig. 4.45 a, b. Spondylitis. Nonspecific x-ray pattern. Differential diagnosis includes staphylococcus, tuberculosis, or brucellosis. The final diagnosis lies on the identification of ancient DNA of the causal microbial agent

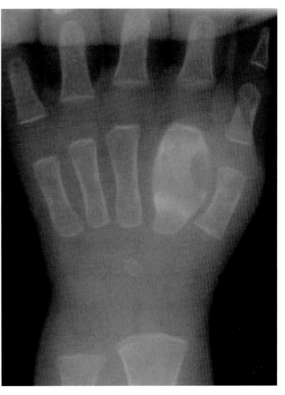

Fig. 4.46. Tuberculous dactylitis or spina ventosa in a child (clinical case, courtesy of Professor Wang)

Table 4.21. Tuberculosis of the skeleton: subtypes

Spine	Limbs
Spondylitis	Osteomyelitis
Spondylodiscitis	Arthritis
	Dactylitis (spina ventosa)

In tuberculous spondylitis, the initial infection can occur at the vertebral body or the posterior elements of the vertebra. Infection of the vertebral body leads to its progressive destruction and collapse. It may also spread through the endplate to destroy the intervertebral disc. Radiological study shows the following patterns: radiolucent lesions of variable size within the vertebral body, narrowing of the disc space, collapse of the vertebral body, and ultimately an acute kyphosis centered on the collapsed vertebra.

The same bone destruction can be seen at the pedicle, lamina, and spinous process. Although erosion of the anterior part of the vertebral body has been suggested as a possible sign of tuberculous spondylitis, it is extremely difficult to distinguish tuberculosis from spinal infection caused by *Staphylococcus*. In the clinical setting, the final diagnosis is established using microbiological data. It is possible that mycobacterial DNA study may aid in the diagnosis of skeletal tuberculosis in dry bone specimens (Herman and Hummel 2003; Zink et al. 2001).

Multiple tubercles are formed in mycobacterial osteomyelitis, which present on x-rays as multiple, relatively well-defined radiolucent lesions in the medullary cavity of the cortex of long tubular bones such the tibia or fibula (Resnick and Kransdorf 2005). Pathological study actually shows foci of osteolysis filled with infectious granulomas called tubercles. They may be surrounded by a thin sclerotic rim and associated with periosteal reactions. They are commonly located in the metaphysis of long bones in children. Indeed, these radiological patterns are not specific. Differential diagnosis includes other types of osteomyelitis and tumors such as metastasis or multiple myeloma.

Tuberculous arthritis characteristically appears as a triad of findings that includes periarticular osteoporosis, gradual narrowing of the joint space, and marginal erosions (Resnick and Kransdorf 2005). The two former signs are not reliable in paleopathology because taphonomic changes can alter the density of the dry bone specimen. Also, joint space is not always preserved due to postmortem changes. Therefore, only marginal erosions should be retained as a sign for tuberculous arthritis. However, marginal erosions are not specific as they may be present in any inflammatory and infectious arthritis. Also, bony ankylosis and bone proliferation are rare in tuberculous arthritis (Rogers and Waldron 1995).

Tuberculous dactylitis (Fig. 4.46) occurs in the short tubular bones of the hands and feet and is much more common in children. X-rays show an expansion of the entire bone called "spina ventosa". When they are bilateral, it should be distinguished from syphilis (Resnick and Kransdorf 2005).

In summary, x-ray findings may suggest skeletal tuberculosis. The gold standard that allows the establishment of tuberculosis is the extraction of microbacterial DNA from skeletal remains (Haas et al. 2000b).

4.74.4
Leprosy

Leprosy is an infectious disease caused by *Mycobacteria leprae* that affects the skin, mucosa, and the peripheral nerves (Milgram 1990; Resnick and Kransdorf 2005). The incubation period is extremely long

and is estimated to be between 3 and 6 years. *M. leprae* infections are divided into four forms: lepromatous, tuberculoid, diphormous, and intermediate (Resnick and Kransdorf 2005). Clinical symptoms include malaise, lethargy, fever, skin manifestations, and adenopathies (Milgram 1990; Resnick and Kransdorf 2005). Musculoskeletal lesions are infection caused by the pathogen and neuropathic joint secondary to nerve damage. Bone infections manifest as periostitis (infection of the periosteum), osteitis (infection of the cortex), and osteomyelitis (infection of spongiosa and bone marrow). Bone infections are located to the face, hands and feet (metaphysis; Figs. 4.47 and 4.48), tibia and fibula (periosteum). The radiological patterns include osteolysis and periosteal reactions, but rarely sclerosis (Table 4.22). Septic arthritis x-ray patterns are those of subacute infection as seen in tuberculosis. As for any bone infections, x-ray patterns are not specific for the pathogen. However, the distribution of bone lesions within the skeleton is key to the diagnosis (Manchester 2002). In the clinical situation, the diagnosis of leprosy is established on the basis of histological changes and the identification of the bacterium.

Histopathological studies have been suggested as an accurate test for determining the diagnosis of leprosy (Schultz and Roberts 2002). The extraction of *M. leprae* DNA is the best evidence with which to establish an unequivocal diagnosis of leprosy in skeletal remains from archeological settings (Haas 2000a).

The second musculoskeletal manifestation of leprosy is neuropathic joint due to denervation, which includes bone resorption and joint destruction, similar to those caused by syphilis, diabetes mellitus, and syringomyelia. This affects the hand, wrist, ankle, and feet joints (Resnick and Kransdorf 2005). Other radiographic findings include osteopenia, atrophy, increased radiodensity, hyperostosis and insufficiency fractures (Cooney et al. 1944; Esguerra-Gomez and Acosta 1948; Faget and Mayoral 1994) (Table 4.22). A comprehensive study of medieval leprosy in Europe had been thoroughly undertaken by Moller-Christensen (1961).

4.7.4.5
Brucellosis

Brucellosis most commonly affects the lumbar spine (Fig. 4.45). It is caused by a bacterium called *Brucella*, which used to be common in cows and was transmitted to humans through the ingestion of infected milk. Radiologically, spondylodiscitis of the spine caused by *Brucella* is not very different from spondylodiscitis due to tuberculosis or *S. aureus*, although some suggests that a "parrot beak" osteophyte at the anterosuperior aspect of the vertebral body may be a characteristic pattern (Resnick and Kansdorf 2005). X-ray signs of spondylodiscitis include disc-space narrowing (not valid in dry specimens), irregularity of the vertebral endplates, and radiolucent areas within the vertebral body or posterior arch. Some sclerosis may accompany the bone destruction. Overall, there are no definite specific radiological signs for Brucello-

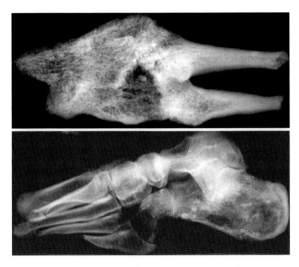

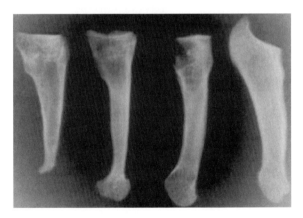

Fig. 4.47 a, b. Leprosy in the foot. Cannington, Dark Age

Fig. 4.48. Leprosy at the toes. Dakhleh, Egypt, Roman Period (courtesy of Dr. El Molto)

Table 4.22. Radiological patterns of leprosy

1.	Osteopenia
2.	Radiolucent areas
3.	Resorption of phalanges
4.	Atrophy
5.	Increased density
6.	Hyperostosis
7.	Insufficiency fractures

sis of the spine. Molecular study with extraction of *Brucella* DNA is the single best test with which to establish a firm diagnosis of spondylodiscitis caused by this bacterium.

4.7.4.6
Paget's Disease

It has been suggested that Paget's disease (Chapman and Nakielny 2003; Milgram 1990; Resnick and Kransdorf 2005) may be caused by an indolent virus infection, but this hypothesis is yet to be confirmed. Paget's disease generally affects the elderly, but may sometimes present as early as 35 years of age. Clinical features vary considerably depending on the location of the lesion. It is asymptomatic in most cases, especially when it affects the sacrum or pelvis. Paget's of the skull leads to deformity and sometimes headaches, hearing impairment, and/or dental malocclusion. Involvement of the spine leads to neurological symptoms due to spinal stenosis, with acute paraplegia occurring in some severe cases. Paget's of the long bones causes bone pain and bowing. Complications include fracture and arthropathies. Groin pain while walking is seen in Paget's of the hip and pelvis. Malignant transformation is seen in less than 1% of cases.

The diagnosis is essentially made radiographically (Resnick and Kransdorf 2005; Roberts and Manchester 2005) (Tables 4.23 and 4.24). The x-ray findings are variable depending on the stage of the disease: active, intermediate, or inactive. In the active stage, bone resorption predominates and occurs in the skull and long bones. In the intermediate stage, there is a mixture of lytic and sclerotic lesions. In the inactive stage, the lesions appear mainly sclerotic (Fig. 4.49). Overall, there will be a softening and deformity of the bone.

4.8
Tumors

4.8.1
Classification of Bone Tumors

In clinical medicine, diagnosis of bone tumors using x-ray is the most exciting and challenging intellectual exercise for a skeletal radiologist. In contrast to "other tumors," when the pathologist has the final say, in bone tumor they always correlate the histologic findings with those provided by x-ray studies in order to prevent misdiagnosis (Table 4.25). The "tumor board," which includes skeletal radiologists, pathologists, orthopedic surgeons, and cancer specialists, is the ideal place to discuss the diagnosis of bone tumors

(Chhem and Ruhli 2004). Although bone tumors are rare in archeological settings, the logical approach for the establishment of the diagnosis of these lesions is worth the description.

Table 4.23. Paget's disease x-ray findings

Lytic	Area of radiolucency with sharp margin in the skull and long bones (focal osteoporosis, bone resorption or fatty transformation of bone marrow)
Sclerotic	Long bones: Thickening of trabeculae, endosteal, cortex, and periosteal reaction involving the epiphysis
	Skull: thickening of both skull tables with "cotton wool" areas
	Spine: ivory vertebra and "picture frame" vertebra with increased AP diameter of the vertebral body
	Pelvis: thickening of the trabeculae and iliopectineal line
Mixed	A combination of the lytic and sclerotic findings

Table 4.24. Ivory vertebra

1.	Metastases
2.	Paget (+ increase AP diameter of vertebral body)
3.	Lymphoma
4.	Infection

Table 4.25. Classification of bone tumors (Lichstenstein 1972)

Cartilage derivation	Benign: osteochondroma, enchondroma, chondroblastoma
	Malignant: chondrosarcoma
Bone derivation	Benign: osteoma, osteoid osteoma, osteoblastoma
	Malignant: osteosarcoma
Connective tissue derivation	Benign: nonossifying fibroma
	Malignant: fibrosarcoma
Hematopoietic origin	Multiple myeloma
	Leukemia
	Lymphoma

4.8.2
Basics of x-ray Interpretation of Bone Tumors

There are three important facts to keep in mind when one interprets x-ray results of bone tumors (Lichstenstein 1972) (Tables 4.26–4.34):

1. The x-ray pattern is not the image of the tumor itself.
2. The x-ray pattern is the image of bone destruction by the tumor and the bone reaction to confine it.
3. The x-ray pattern is the result of imbalance between bone destruction caused by tumor and adjacent bone formation

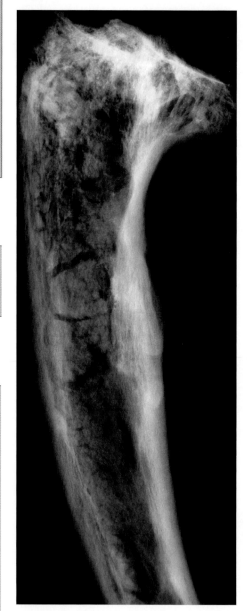

Fig. 4.49. Paget's disease of the left tibia

4.8.2.1
Periosteal Reactions

The periosteum is a cellular layer that demarcates bone from the surrounding soft tissue (Ragsdale et al. 1981). It is composed of two layers: an inner cambium layer and a dense outer fibrous layer. The inner end-osteal surface is attached to the underlying cortical bone by Sharpey fibers, which become increasingly

Table 4.26. Common radiologically "aggressive-looking" lesions

1.	Malignant primary bone tumors
2.	Metastasis (malignant secondary tumors)
3.	Osteomyelitis
4.	Eosinophilic granuloma

Table 4.27. Bone destruction and formation in skull

Bone destruction in skull	Metastasis
	Multiple myeloma
	Paget
	Infection
	Fibrous dysplasia
	Eosinophilic granuloma
	Trauma
	Trepanation
Bone formation in skull	Osteoma
	Osteochondroma (ear)
	Paget
	Meningioma
Destruction of facial sinuses	Leprosy
	Trauma (mummification)
	Fibrous dysplasia
	Tumors

Table 4.29. Solitary lytic bone lesions: x-ray patterns, checklist

X-ray patterns	Types
Location: epiphysis, metaphysis, diaphysis	I A, B, C, II and III
Cortical, trabecular bone or both	
Shape, size, margins, transition zone	
Solitary or small and multiple	
Mineralized matrix (osteoid or cartilaginous)	
Cortical destruction or "expansion"	
Periosteal reactions or not	

Table 4.30. Type IA/IB lesions

1.	Enchondroma
2.	Fibrous dysplasia
3.	Simple bone cyst
4.	Nonossifying fibroma
5.	Eosinophilic granuloma

Table 4.31. Type IC, II lesions

1.	Osteosarcoma
2.	Ewing tumor
3.	Chondrosarcoma
4.	Giant cell tumor
5.	Lytic metastasis
6.	Multiple myeloma
7.	Osteomyelitis
8.	Eosinophilic granuloma

Table 4.28. Classification of bone tumors according to biological activity

Internal margins	Periosteal reactions	Biologic activity	Diagnosis
Geographic			
IA	Solid	Slow nonaggressive	Enchondroma, chondroblastoma, chondro-myxoid fibroma, osteoblastoma, bone cyst, fibrous dysplasia
IB	Solid	Slow nonaggressive	Same as IA, giant cell tumor
IC	Interrupted Shell	Intermediate	Same as IA and IB undergoing malignant transformation, bone sarcomas
Moth-eaten			
II	Interrupted Shell	Intermediate	Small, round-cell lesions, bone sarcomas
Permeative			
III	Lamellated	fast	Same as II

Table 4.32. Type III lesions

1.	Ewing tumor
2.	Chondrosarcoma
3.	Osteomyelitis

Table 4.33. Bone tumors: diagnosis based on prevalence, number, and location within individual bone

Two most common secondary tumors	
Metastases: most common	
Multiple myeloma: second most common	
Three most common primary bone tumors	
Osteosarcoma: most common	
Ewing tumor: second most common	
Chondrosarcoma: third most common	
Most common site within individual bones	
Epiphyseal lesions	Chondroblastoma
	Giant cell tumor
Metaphyseal lesions	Osteochondroma (exostosis)
	Osteosarcoma
	Ewing
	Chondrosarcoma
	Osteomyelitis
	Eosinophilic granuloma
	Fibrous dysplasia
	Simple bone cyst
	Aneurysmal bone cyst
Diaphyseal lesions	Ewing
	Chondrosarcoma
	Osteomyelitis

adherent as the skeleton matures. The periosteum of adult bone is only a few cells thick and is predominantly fibrous.

Many types of irritation to bone, including benign and malignant tumors, trauma, inflammation, or infection, will cause the periosteum to react by laying new bone in a characteristic pattern. These periosteal reactions can be continuous or interrupted (Fig. 4.50). The former often accompanies a benign tumor or fracture healing. These nonaggressive processes give the periosteum adequate time to lay down continuous new bone, which may appear thick or wavy. When the cortex is interrupted, the bone contour may appear widened. This is called a "shell" pattern. Interrupted periosteal reactions usually indicate a more aggressive lesion such as a malignant tumor or infection. The speed of progression does not provide the periosteum adequate time for consolidation. The periosteum

Table 4.34. Bone tumor: most common skeletal sites

1.	Nonossifying fibroma: metadiaphyseal regions of the tibia and distal femur
2.	Simple bone cyst: proximal femur/humerus
3.	Enchondroma: small bones of hands/feet
4.	Osteochondroma: distal femur or proximal humerus
5.	Giant cell tumor: distal femur, proximal tibia or humerus
6.	Osteosarcoma: "close to knee" (proximal tibia and distal femur); "away from elbow" (proximal humerus and distal radius)
7.	Fibrous dysplasia: femur, tibia, ribs

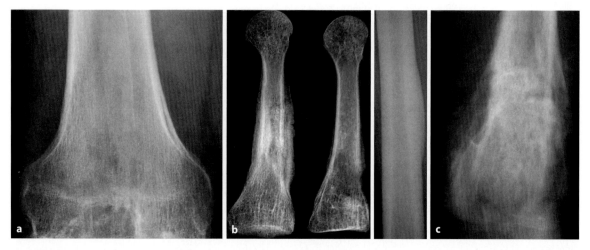

Fig. 4.50 a–d. Types of periosteal reaction. **a** Unilamellar and continuous. **b** Continuous. **c** Solid and continuous. **d** Disrupted "Codman's triangle"

reacts by inducing bone formation perpendicular to the surface of the cortex, appearing spiculated ("hair-on-end," "sunburst"), or the bone formation may be parallel to the surface with a lamellated appearance ("onion-skin" when the lamellated reactions are continuous, "Codman angle" when they are interrupted).

It must be kept in mind that there are always exceptions; some aggressive processes will induce continuous periosteal reactions, and benign processes may at times produce an interrupted appearance. Another important caveat is that continuous and interrupted periosteal reactions are not mutually exclusive; both types may occur in a single complex lesion. The most important concept to understand from this section is that the term "aggressive" lesion is only an expression of the lesion's biological activity, whether the lesion is a tumor, an infection or a trauma. The terms "benign" and "malignant" are defined histologically. Hence, an aggressive type of periosteal reaction is a radiological term, and does not necessarily equate with malignant bone tumor.

4.8.2.2
Internal Margins

Lytic lesions of osseous structures are those that cause a focal loss of bone and therefore appear radiolucent relative to normal bone. The border surrounding a lytic lesion is called the internal margin, or zone of transition (Madewell et al. 1981). Internal margins are important because, like periosteal reactions, they provide information about the aggressiveness of the lesion. Well-defined lesions tend to be nonaggressive, whereas those with indistinct borders are more likely to be aggressive. Again, there are always exceptions; some entities, such as bone metastases, can have various appearances.

Internal margins of lytic lesions are classically described in terms of three destruction patterns: geographic, moth-eaten, or permeative (Fig. 4.51; Tables 4.31–4.32). Geographic lesions (type I) are solitary and lytic. They may have a narrow zone of transition, either sclerotic (IA) or absent (IB), indicating slow growth. Geographic lesions with ill-defined margins (IC) indicate a more rapid growth pattern and are therefore more likely to be aggressive, with the understanding that the term "aggressive" as used here characterizes only the biological activity. Malignant bone tumors, bone infection, and eosinophilic granulomas can appear aggressive. As mentioned previously in the section 4.8.2.1 on periosteal reactions, an important point is that not all aggressive, solitary bone lesions are malignant tumors. Moth-eaten lesions (type II), as the name implies, appear as multiple, small, rounded lesions of variable size that arise separately, rather than from the edge of a main central lesion and then coalesce. They occur in both cancellous and cortical bones. Permeative lesions (type III) are also multip-

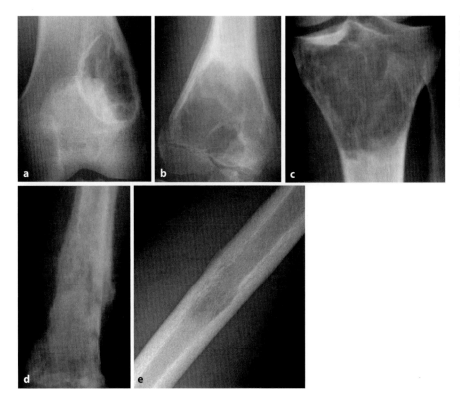

Fig. 4.51 a–e. Types of lytic bone lesions (clinical cases). **a** Lytic lesion type IA. **b** Lytic lesion type IB. **c** Lytic lesion type IC. **d** Lytic lesion type II. **e** Lytic lesion type III

le oval or linear lucent lesions, but occur mostly in the cortex. Both types II and III are characteristic of aggressive processes. Distinguishing between them is not essential because they have the same differential diagnoses. It is far more important to distinguish between less aggressive (IA, IB), intermediate (IC), and more aggressive (II, III) lesions.

4.8.2.3 Matrix Patterns

Bone tumors are composed of a collection of collagen fibers and crystalline salts called the matrix (Sweet et al. 1981). There are many different types of matrices, but only two can be clearly identified on radiographs: osteoid and chondroid matrices (Fig. 4.52). The cellular difference between the two pertains to how readily the crystalline salts precipitate on the collagen fibers. Osteoid matrix tumors include osteoid osteoma, osteoblastoma, bone island, and osteosarcoma. These tumors usually mineralize in a way that results in radiographic patterns described as dense, homogenous, or cloudlike. Chondroid matrix tumors include enchondroma, chondroblastoma, chondrosarcoma, and chondromyxoid fibroma. These tumors may not mineralize, or will do so with a radiographic pattern of arcs or circles. Other osseous lesions are said to have a cellular matrix, neither osteoid nor chondroid. Fibrous matrix cannot be distinguished as such on x-ray, but rather appears as a nonspecific radiolucent area. Fibrous dysplasia is one such example, typically having a diffuse ground-glass appearance.

The radiographic examination of a solitary lesion should follow the checklist given in Table 4.29. The systematic analysis of the basic x-ray patterns of solitary lytic lesions is an extremely important step in the logical approach to the differential diagnosis. In addition to the radiographic features, the type of lytic lesions (IA, IA, IC, II, and III) and the location and number of the lesions within the bone are key to the differential diagnosis of solitary bone lesions (Fig. 4.53). By correlating the radiographic features with the age at death of the specimen (age of patient in clinical cases), one can narrow the gamut of diagnoses to one, two, or at the most three probable bone diseases. This is the way skeletal radiologists cognitively process their approach to bony lesions in daily practice. Clinical radiologists are, however, fortunate in being able to incorporate other data, unavailable to paleoradiologists, such as clinical examination, clinical history, and laboratory data to accurately establish the final diagnosis. Despite the wealth of clinical data, and the availability of advanced imaging tests such as CT and MRI, some lesions may require further investigation with surgical biopsy, and histological or bacterial evaluation to reach the right diagnosis. All of this demonstrates the challenges faced by paleoradiologists and paleopathologists in establishing the final diagnosis of paleopathological specimens without the advantages enjoyed by clinical radiologists. While the advice of a skeletal radiologist in the interpretation of x-rays of the pathological specimen is recommended, this chapter is designed to assist paleopathologists in

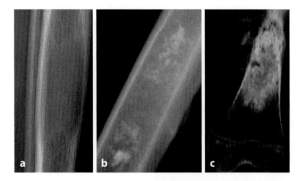

Fig. 4.52 a, b, c. Types of tumor matrix. **a** fibrous. **b** chondroid. **c** osteoid

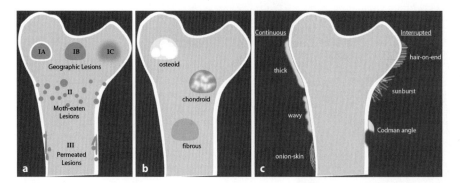

Fig. 4.53 a, b, c. Summary of x-ray patterns of lytic bone lesions. **a, b** Types IA and IB indicate nonaggressive bone destruction, whereas types IC, II, and III indicate an aggressive process. "Aggressiveness" reflects only the biological activity of the lesion, not the histologic nature of the lesion. Nonaggressive lesions may represent a malignant tumor, while an aggressive lesion is not always a malignant tumor (it can represent infection or eosinophilic granuloma). **c** Continuous (nonaggressive) and interrupted (aggressive) periosteal reactions

their efforts to evaluate bony lesions when a skeletal radiologist is not available.

4.8.3
Common Tumors

The purpose of this section is not to provide a thorough review of skeletal tumors as diagnosed by x-ray study, but instead to describe the clinical and radiological profiles of common skeletal lesions as reported in the physical anthropology and paleopathology literature (Chapman and Nakielny 2003; Lichtenstein 1972; Resnick and Kransdorf 2005). Readers interested in further expanding the scope and depth of their knowledge in skeletal tumors are advised to read the relevant textbooks listed in the references.

4.8.3.1
Osteochondroma or Exostosis

They are solitary in most cases (Fig. 4.54). Multiple exostosis is a rare dominant autosomal hereditary disease (Fig. 4.55). Malignant transformation is rare in solitary exostosis, but may occur in multiple hereditary exostosis. Exostoses cease development when the epiphyseal plates close. The tip of an exostosis points away from the growth plate as the result of muscle pull during skeletal growth and development. The cortex and medulla of the osteochondroma are continuous with the host bone, as demonstrated on x-rays films. While there is no serious disability

associated with osteochondromas because they are slow-growing tumors, they can sometimes be painful as they enlarge, causing neurological symptoms when nerves are compressed, and difficulty walking when the tumor arises from the ankle and/or foot. In clinical cases, 80% occur in patients less than 21 years old. There is no gender predominance. Differential diag-

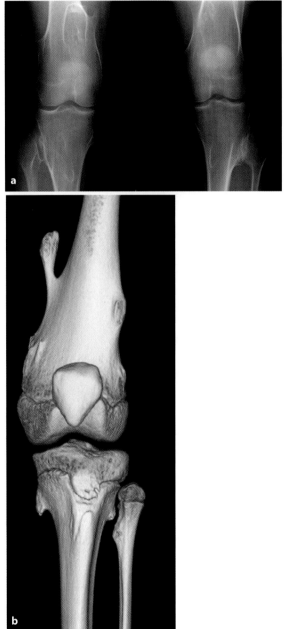

Fig. 4.55 a, b. a Familial hereditary multiple exostosis. Clinical case. **b** 3D CT: familial multiple hereditary exostosis. Clinical case. "Beautiful 3D" adds no value to the diagnosis already achieved using x-ray. Courtesy of GE

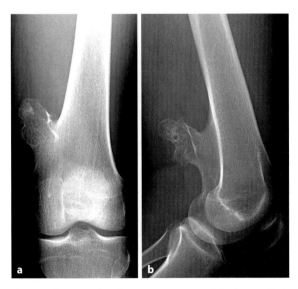

Fig. 4.54 a, b. Osteochondroma or exostosis of the femur: clinical case

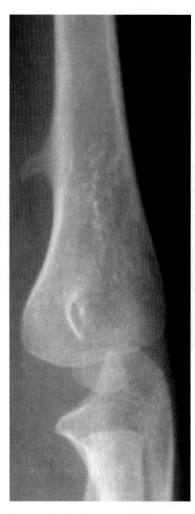

Fig. 4.56. "Feline" spur: normal variant, simulating osteochondroma (exostosis) or ossified enthesopathy

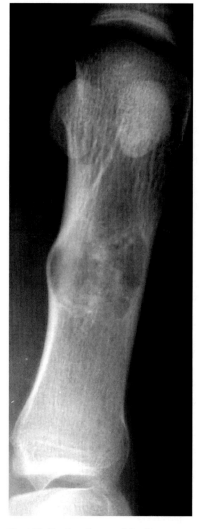

Fig. 4.57. Enchondroma of the first metatarsal. Clinical case

noses include ossified enthesopathy and feline spur, a normal anatomical variant (Fig. 4.56).

4.8.3.2
Enchondroma

Enchondroma is a well-defined, nonaggressive, osteolytic medullary lesion with a lobulated contour with calcifications of the matrix (Fig. 4.57) that occurs commonly in the hands and feet. Large enchondromas cause endosteal erosion, cortical expansion, and thickening. A pathologic fracture may complicate large enchondroma (Resnick and Kransdorf 2005). There is no gender predilection, and endochondroma usually occurs at between 10 and 50 years of age. These lesions are asymptomatic and rarely painful. They are often discovered incidentally in clinical situations where x-ray examination was performed for trauma.

4.8.3.3
Osteosarcoma

Osteosarcoma is a malignant tumor of the bone developing most often in males between 10 and 25 years of age, with a second peak at around 60 years. Clinical features include months of pain and swelling, weight loss, anemia, and/or pathologic fracture. Osteosarcoma may occur anywhere in the long or flat bones, and without treatment, lung metastasis occurs in 80% of cases within 3 years, leading rapidly to death. Pneumothorax may occur spontaneously and may be the initial manifestation of lung metastasis. Clinically, patients may present with acute chest pain and/or shortness of breath. Radiographic patterns include mixed osteolysis and sclerosis. Osteolysis occurs at the metaphysis of long bones and extends to the cortex, then produces an aggressive type of peri-

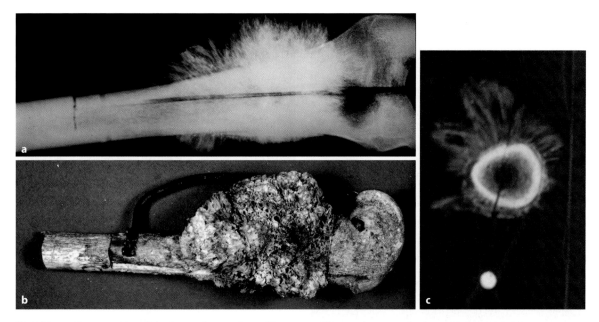

Fig. 4.58 a–c. Osteosarcoma of the femur: dry specimen. Dupuytren Museum (courtesy Professor Vacher-Lavenu)

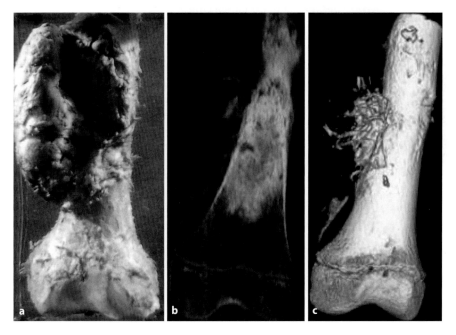

Fig. 4.59 a–c. Osteosarcoma of the femur: a wet specimen, b 2D CT, c 3D CT. the "Beautiful 3D CT" does not add any value to the diagnosis already established by 2D CT

osteal reaction such as Codman's triangle (Resnick and Kransdorf 2005) (Figs. 4.58 and 4.59).

4.8.3.4
Paraosteal Sarcoma

Paraosteal sarcoma is a malignant bone tumor that originates on the external surface of the bone. It is a very rare tumor. The mean age of distribution is ap-proximately 25 years old. Clinical symptoms include swelling, a mass, a dull aching pain, or local tenderness. X-rays show an irregular, lobulated, dense tumor developing around a long bone. The most common location is around the knee region. Other locations are the femur, tibia, humerus, and fibula. Radiologically, this rare malignant tumor must be distinguished from heterotopic bone formation (myositis ossificans; Huvos 1991) (Fig. 4.60).

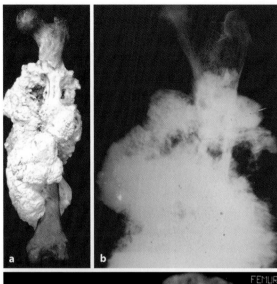

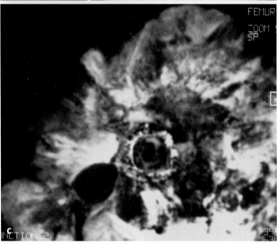

Fig. 4.60 a–c. Paraosteal sarcoma of the femur: dry specimen. Dupuytren Museum (courtesy of Professor Vacher-Lavenu)

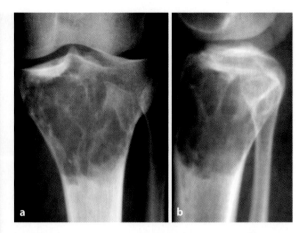

Fig. 4.61 a, b. Giant cell tumor of the tibia. Clinical case

that are vertical to the cortex (Resnick and Kransdorf 2005). In clinical situations, a significant periosseous soft-tissue mass is usually present. This sign is irrelevant in paleoradiology, as soft tissues are long gone in dry bone specimens.

4.8.3.6
Chondrosarcoma

This is an osteolytic lesion with endosteal erosion, cortical thickening, and periosteal reaction. It occurs most commonly in the humerus and femur of males between 30 and 60 years of age, but the lesion can typically take several years to clinically manifest. Symptoms include dull aching local pain with episodes of exacerbation occurring over a 6-months to several-years period. The matrix contains scattered "ring and arc," "popcorn," or "dot and comma" calcifications (Chapman and Nakielny 2003; Resnick and Kransdorf 2005).

4.8.3.7
Giant Cell Tumor

It is an osteolytic lesion of the epiphysis that may extend to the metaphysis and involves the subchondral bone. The lytic lesion contains a delicate trabecular pattern. The margin is either well defined, without any peripheral sclerosis (Type IB) or ill defined (Type IC) (Fig. 4.61). They occur in patients aged 20–40 years, when the growth plate is closed (Resnick and Kransdorf 2005).

4.8.3.8
Osteoma

Osteoma is a benign bone tumor that is characterized by bony excrescences usually arising in membranous

4.8.3.5
Ewing's Sarcoma

This is an aggressive tumor of the bone that involves the tubular and innominate bones as well as the spine. It is most common in males between 10 and 25 years of age. Pain is the most common symptom, typically lasting for months to 1 year before treatment is sought in clinical cases, because the pain has become increasingly severe and persistent. Fever and anemia may also be associated features, and symptoms can vary depending on the tumor location. For example, there is stiffness and pain when the tumor arises near the hip joint. X-rays patterns include a moth-eaten osteolysis of the trabecular bone with poorly defined and variable cortex and periosteal reactions. They appear as laminated, "onion-skin," or multiple strands

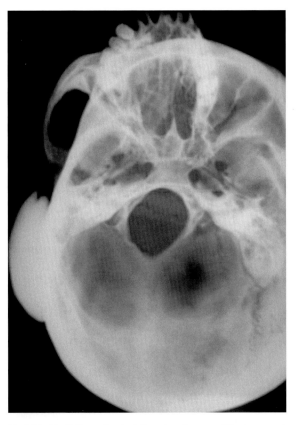

Fig. 4.62. Skull from Ancient Egypt: osteoma of the temporal bone

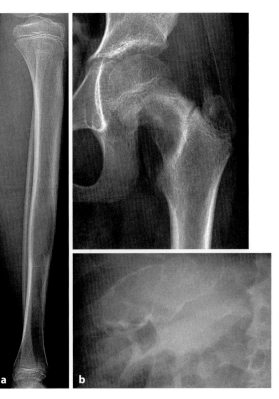

Fig. 4.63 a, b. **a** Fibrous dysplasia of the tibia. Clinical case. **b** Fibrous dysplasia of the femoral neck. Clinical case. c Fibrous dysplasia of the rib. Clinical case

bones. They have well-defined margins and may be sessile or pedunculated. Most osteomas are painless, slowly enlarging hard lumps that can arise from the orbit, the vault of the skull (Fig. 4.62), and occasionally around the external auditory canal. X-rays show a well-defined sclerotic margin (Huvos 1991).

4.8.3.9
Fibrous Dysplasia

Fibrous dysplasia is not a tumor. It is included in this section because its x-ray pattern simulates a nonaggressive bone tumor (Table 4.35). Fibrous dysplasia is described as bone lesions of unknown etiology by Jaffe and Lichtenstein (Lichtenstein 1972; Milgram 1990) that may affect a single or multiple bones. Extraskeletal manifestations may also occur. Histologic studies show a predominantly fibrous matrix separating the osseous trabeculae. Fibrous dysplasia is usually asymptomatic; becoming painful only when there is a stress or complete fracture through the tumor. Bowing and deformity of the extremities are common in advanced cases. The most common finding is the so called "ground-glass appearance" on radiography (Fig. 4.63).

Table 4.35. Fibrous dysplasia

1.	Geographic lesion with the sclerotic margin (IA)
2.	Oval shaped
3.	Endosteal resorption
4.	No cortical rupture, no periosteal reactions
5.	Metaphyseal or diaphyseal
6.	Bone deformity ("Shepherd's crook of the proximal femur/bubbly appearance of the rib)
7.	Leontiasis ossea (deformity of the facial bones)
8.	Mixed type lesion of the skull

4.8.3.10
Simple Bone Cyst

This is the most common benign "tumor" of the appendicular skeleton in children. It is most often asymptomatic and may only become painful when a pathologic fracture occurs. The cyst develops at the metaphysis of long bones with more than 50% at the proximal humerus. The other sites of simple bone cysts are the femur, tibia, calcaneum, or pelvis. X-ray studies show a radiolucent lesion with fine skeletal margins in contact with the growth plate. The cyst may contain a few septae. In skeletal remains, the fluid content of the cyst may dry up. In that situation, the cyst itself appears completely radiolucent and displays a gaseous density (Resnick and Kransdorf 2005).

4.8.3.11
Aneurysmal Bone Cyst

This is a multiloculated cystic cavity containing blood. Aneurysmal bone cysts are either primary or secondary to a benign and malignant tumor. Eighty percent of aneurysmal bone cysts occur before the age of 20 years. In clinical cases, a history of trauma is present in three-quarters of the cases. The cyst develops at the metaphysis of the long bone, at the posterior arch of the vertebrae, or in the pelvic bones. X-ray patterns include a radiolucent lesion containing septae. It is commonly expansile, leading to a significant thinning of the cortical bone (Fig. 4.64). In some cases aneurysmal bone cysts may extend to the adjacent soft tissues with aggressive periosteal reaction that may simulate a malignant tumor (Biesecker et al. 1970).

4.8.3.12
Vertebral Hemangioma

This is a benign neoplasm formed by proliferating blood vessels (Aufderheide and Rodriguez-Martin 1998; Ortner 2003). It is most common in the thoracic spine and is present in 11% of spinal autopsy series (Schmorl and Junghans 1971). It affects individuals after the fourth decade of life and is more common among females than males (Pastushyn et al. 1998). Hemangiomas are usually asymptomatic. Exceptionally, pathologic fracture can occur and may lead to cord compression. In these cases, the entire vertebral body is involved with extension to the posterior arch through the pedicles (Laredo et al. 1986). In dry specimens, x-ray patterns of hemangioma include: a focal area of coarsening of the vertical trabeculae within the vertebral body, expansion of the cortex is present when the hemangioma develops beyond the margins of the vertebral body. The so-called "honeycomb" or "corduroy" pattern is enhanced by the replacement of bone marrow by ambient air that has infiltrated the interstices, separating vertical trabeculae (Fig. 4.65).

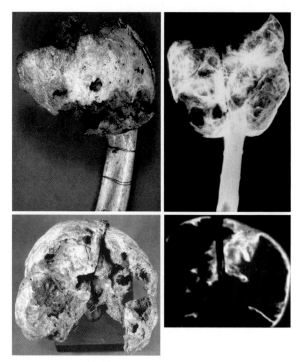

Fig. 4.64 a–d. Aneurysmal bone cyst: dry specimen. Dupuytren Museum (courtesy of Professor Vacher-Lavenu)

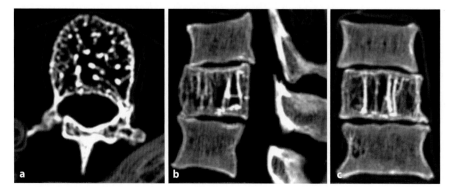

Fig. 4.65 a–c. Vertebral hemangioma T12/L1 (a axial view of T12). Roman Period (circa 150–400 A.D.), Dakheh Oasis, Egypt (courtesy of Dr. El Molto)

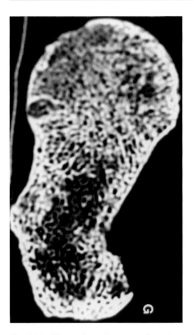

Fig. 4.66. Heniation pit: normal variant. Samrong Sen, prehistoric Cambodia

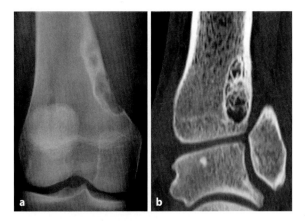

Fig. 4.67 a, b. **a** Nonossifying fibroma of femur. Pathognomonic on x-rays. Clinical case. **b** Nonossifying fibroma: CT

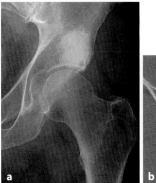
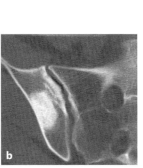

Fig. 4.68 a, b. **a** Giant bone island, simulating a metastasis. **b** Giant bone island of iliac bone: CT

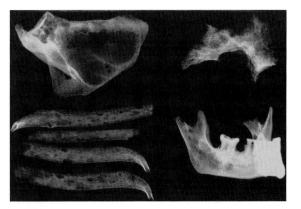

Fig. 4.69. Metastasis versus multiple myeloma. Prehistoric Amerindian from the Berkeley collection

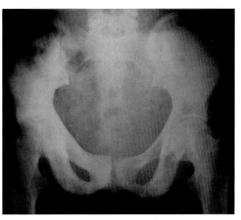

Fig. 4.70. Sclerotic skeletal metastases from prostate cancer (courtesy of Dr. Friedman)

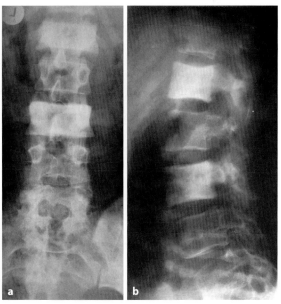

Fig. 4.71 a, b. Sclerotic metastases of the spine ("ivory vertebra"): x-ray, anteroposterior view. Clinical case

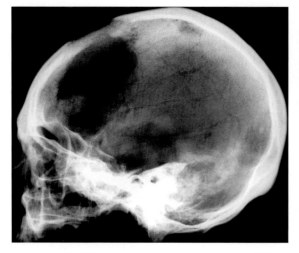

Fig. 4.72. Skull x-ray: prehistoric Amerindian from the Berkeley collection. Metastases versus multiple myeloma

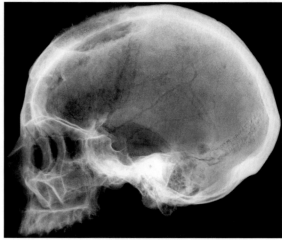

Fig. 4.73. Skull x-ray: Ancient Egypt. "Aggressive" tumor. No specific histologic diagnosis is possible with this x-ray study alone

4.8.3.13
Pseudotumor

These are characteristic and must be differentiated from authentic tumors (Figs. 4.66–4.68). In any doubt, readers must systematically consult the "Keats" (Keats 1988).

4.8.3.14
Metastases Versus Multiple Myeloma

Skeletal metastases are the most common secondary bone tumor. They occur in the advanced stage of a primary cancer from other regions of the body. They can be either sclerotic or lytic (Figs. 4.69–4.72). The main criterion for the establishment of diagnosis of skeletal metastases is their multiplicity. Multiple lytic metastases must be differentiated from multiple myeloma, which is a malignant tumor of plasma cells. A description of clinical and radiological patterns of skeletal metastases and multiple myeloma is beyond the scope of this chapter. As a rule, multiple lytic lesions of the skeleton suggest either a metastases or multiple myeloma. Finally, not all dry-bone tumors can be diagnosed based on x-rays – beyond their aggressiveness (Fig. 4.73). Care must be taken to exclude taphonomic changes like insect bites (Fig. 4.74).

4.9
Metabolic, Endocrine, Ecosystem Diseases, and Anemias

This section discusses the paleoradiological method in the diagnosis of selected diseases that have been

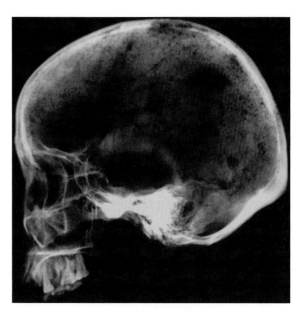

Fig. 4.74. Taphonomic Nubian skull: insect bites

Table 4.36. Skull: abnormal size and shape

1. Craniosynostosis (plagiocephaly, brachycephaly, dolichocephaly, scaphocephaly, trigonocephaly)
2. Platybasia: rickets, osetomalacia, Paget, fibrous dysplasia, osteogenesis imperfecta
3. Metabolic: rickets, hyperthyroidism, hypophosphatasia
4. Congenital: Morquio's and Hurler's syndromes
5. Hematological: thalassemia, sickle cell
6. "Cultural": Maya's skull

detected in skeletal remains (Table 4.36). The description of the pathogenesis of each of these diseases is beyond the scope of this book.

4.9.1
Congenital Skeletal Diseases

The diagnosis of congenital skeletal lesions is usually established exclusively by plain-film radiography. Some diagnoses are straightforward: for example, tarsal or carpal coalition, or abnormal skull shape, while others are more difficult. The diagnosis of congenital skeletal disease is challenging even for pediatric radiologists with experience in skeletal dysplasia.

The analysis of x-ray studies starts with a methodic evaluation of basic radiological findings, such as the shape, size, and the exact location of the lesions (i.e., epiphysis, diaphysis, metaphysis). The pattern of the distribution gives clues to the diagnosis. These features are combined and then correlated with the "catalogue" of skeletal dysplasia published in the "Radiology of Syndromes, Metabolic Disorders, and Skeletal Dysplasias" book (Lachman and Hooshang Taybi 1996). If the final diagnosis cannot be established after this careful approach, then the final option would be to contact members of the International Skeletal Dysplasia Society (http://www.isds.ch/ISDSframes.html) for advice.

4.9.2
Osteoporosis

Osteoporosis is a skeletal disease that leads to decreased bone mass. Early osteoporosis starts at the spine, pelvis, sternum and ribs, while in advanced osteoporosis abnormal findings are seen in the extremities. Paleopathologists consider the diagnosis of osteoporosis when the bones are lighter in terms of weight compared to similar specimens from the same archeological setting. Visual observation of the specimen is normal in most cases. Only a cut surface shows

Table 4.37 X-rays of osteoporosis

1.	Decrease bone density (consider taphonomic changes)
2.	Thinning and rarefaction of bone trabeculae
3.	Thinning of cortical bone (second metacarpal)
4.	Anterior wedging of the vertebral body
5.	"Fish" vertebra (biconcave vertebral body)
6.	Insufficiency fracture (stress fracture)

a thinning of the cortex and rarefaction of trabecular bone.

Radiological findings vary depending of the anatomical location (Table 4.37). The main x-ray pattern is a decrease in bone density, described as osteopenia in the radiological literature, to characterize any skeletal disturbance that has led to decreasing bone mass (osteoporosis) and/or demineralization (osteomalacia). The second common pattern is a rarefaction of bone trabeculae, which become thinner (Helms 2005; Ortner 2003; Resnick and Kransdorf 2005; Steinbock 1976a).

In the spine, severe osteoporosis leads to a compression fracture of the vertebral body that may affect the endplates and/or the anterior aspect of the vertebral body. Anterior wedging and "fish vertebra" are the two most common radiological features associated with vertebral compression fracture. Fracture of the vertebral body in the setting of osteoporosis occurs most commonly at the midthoracic or thoracolumbar spine (Fig. 4.2). For any fracture cephalad to T7, other causes such as metastasis or multiple myeloma should considered in the differential diagnosis (Resnick and Kransdorf 2005). Osteoporosis of the long bones demonstrates thinning of the cortex and trabecular bones. Also, resorption of the endosteal surface of the cortex leads to a widening of the medullary cavity.

Biparietal thinning is not a reliable sign of osteoporosis, because it is also associated with normal developmental growth in young individuals (Ortner 2003). Insufficiency fracture is a stress fracture that occurs in a pathologically weakened bone and is a typical complication of severe osteoporosis. A quantitative method to assess osteoporosis is to perform an x-ray study of the hands to measure the combined thickness of the cortex at the level of the midshaft of the second metacarpal. The same radiological findings used clinically may be applied to specimens from the archeological record, assuming that taphonomic changes are taken into consideration. For example, postmortem changes attributed to burials and soil conditions may affect the density of the bones (Roberts and Manchester 2005; Steinbock 1976a).

In summary, osteopenia is not a reliable radiological sign for osteoporosis, as many confounding factors may affect bone density; however, thinning of the cortex of long bones and compression fractures of osteopenic vertebral bodies are often associated with osteopenia and/or osteoporosis.

4.9.3
Osteomalacia and Rickets

Osteomalacia is a metabolic bone disease where the osteoid matrix fails to mineralize properly due to an

Table 4.38. Causes of osteomalacia

1.	Vitamin D deficit (nutritional, deficit of sunlight)
2.	Intestinal malabsorption
3.	Chronic renal disease

Table 4.39. X-rays of osteomalacia

1.	Intracortical tunneling
2.	Blurred and coarsened trabecular pattern
3.	Blurring of vertebral end plates
4.	Generalized osteopenia
5.	Pseudofracture (femoral neck, pubis, axillary border of the scapula, radius, and ulna)
6.	True fracture
7.	Bone deformity

Table 4.40. X-rays of clinical rickets

1.	Osteopenia at metaphysis
2.	Enlargement of growth plate
3.	Cupping of metaphysis
4.	Bone spurs at the metaphysis
5.	Cupping at the anterior aspect of the ribs

Table 4.41. X-rays of hyperparathyroidism

1.	Osteopenia
2.	Subperiosteal resorption (radial side of metacarpals, sacroiliac joints, distal clavicle)
3.	Brown tumors

abnormality of vitamin D metabolism (Milgram 1990; Ortner 2003; Resnick and Kransdorf 2005; Steinbock 1976a) (Table 4.38). Rickets is the same basic disorder but occurs in the growing child (Milgram 1990). Radiological findings of osteomalacia vary depending on the stage and severity of the disease and include intracortical tunneling, blurred and coarsened trabecular patterns, blurring of vertebral end plates, and generalized osteopenia (Table 4.39). Complications are the result of severe bone weakening and include bone deformity and pseudofracture (of the femoral neck, pubis, axillary border of scapula, radius, and/or ulnar bones; Aegerter 1975; Griffith 1987). Rickets is termed "a disease of civilization," where the "crowd-

ed, overhanging houses of the city would block out any sunlight that managed to penetrate the barrier of industrial smoke" (Roberts and Manchester 2005). Radiographic signs of rickets are: osteopenia at the metaphysis, enlargement of the growth plate, cupping of the metaphysis, bone spurs at the metaphysis, and cupping at the anterior aspect of the ribs (Tables 4.40 and 4.41).

X-rays of the wrist in a child with typical radiological signs of rickets include widening of the growth plate (not a reliable sign in dry specimens), fraying and cupping of the metaphysis, bone spurs extending from the metaphysis (Fig. 4.75).

4.9.4
Harris Lines

Harris lines of arrested growth are extremely well known to paleopathologists. They are thin, dense lines running transversely across the shaft of long bones, most commonly involving the femur, tibia, and radius (Harris 1930; Roberts and Manchester 2005). Harris lines are not specific for any particular disease. In most cases they reflect nutritional deficiencies and childhood diseases.

4.9.5
Avascular Necrosis–Bone Infarcts

Necrosis of bone is the result of the obliteration of the blood supply to the bone (Milgram 1990; Resnick and Kransdorf 2005). Cell necrosis in bone cannot be demonstrated by x-ray study; however, the reaction of bone to that ischemic process is manifested on radiographs. Bone necrosis can be induced by a multitude of causes that include trauma, hemoglobinopathies, pancreatitis, dysbaric conditions, alcohol consumption, and Gaucher's disease. Necrosis can occur in the epiphysis, metaphysis, and diaphysis. The radiological findings depend on the stage of the disease. In the early stage, the x-ray study is completely normal. As the disease progresses, the x-ray shows osteopenia, cystic areas, and bone sclerosis, particularly at the femoral head, which is the most common site of bone necrosis. In the advanced stage, there is a collapse of the subchondral bone of the femoral head separated from the underlying bone by a crescent-shaped radiolucency. In the final stage, there is a narrowing of the joint space and development of secondary osteoarthritis. Necrosis of the diaphysis appears as a shell-like calcification of the geographic necrotic area. It should be differentiated from enchondroma, where calcifications are clustered in the center of the lesion.

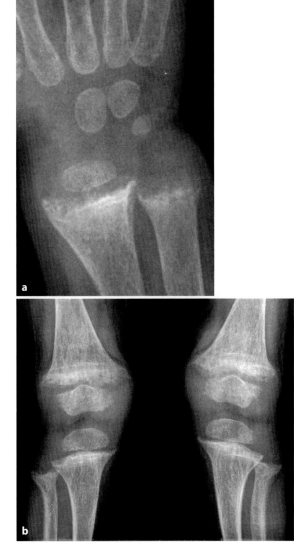

Fig. 4.75 a, b. **a** Rickets at the wrist. X-rays of the wrist in a child with typical radiological signs of rickets, which include widening of the growth plate (not a reliable sign in a dry specimen), fraying and cupping of the metaphysis, and bone spur extending from the metaphysis. Clinical case (courtesy of Dr. Oudjhane). **b** Rickets at the knees. X-rays are similar to those described at the wrist. Clinical case (courtesy of Dr. Oudjhane)

Table 4.42. Possible causes of cribra orbitalia

Anemia
Sickle cell disease
Thalassemia
Hereditary spherocytosis
Iron deficiency
Vitamin B12, vitamin C deficiency
Osteitis
Taphonomic changes

Table 4.43. "Hair-on-end"

Hemolytic anemia	Thalassemia
	Sickle cell anemia
	Spherocytosis
	Elliptocytosis
Tumors	Hemangioma
	Meningioma
	Metastasis

lia is not specific for anemia, as it may be associated with osteoporosis, inflammation of the orbital roof, or post mortem erosions (Walper et al. 2004). Cribra orbitalia has been associated with other types of anemia including sickle cell disease, thalassemia, and hereditary spherocytosis (Caffey 1937) (Table 4.42). Deficiencies in iron, vitamin B12, and vitamin C have also been suggested as causes of cribra orbitalia (Ortner 2003; Stuart-Macadam 1992). In physical anthropological studies, the diagnosis of cribra orbitalia is established mainly on visual inspection of the dry skull, and radiological study has rarely been used to assist in the diagnosis. The pathogenesis of cribra orbitalia has been addressed thoroughly in the paleopathology literature, but there is no current consensus. Further investigation including imaging (especially CT and micro-CT), DNA, and histopathology may be helpful to solve this issue. X-ray studies of chronic hemolytic anemia include a thickening of the skull, a "hair-on-end" pattern, erosion of the outer table of the skull, small facial sinuses (Aksoy et al. 1966; Reimann et al. 1975; Sebes and Diggs 1979; Steinbock 1976b) (Table 4.43; Figs. 4.76 and 4.77). A recent study did report that CT was useful in displaying the bony changes seen in cribra orbitalia (Exner et al. 2004), but the study failed to address the challenging issue of determining the exact cause of those bony changes at the orbital roof. In the future, data obtained through micro-CT study may shed some light on both the pathology and the pathogenesis of cribra orbitalia.

4.9.6 Anemias

Chronic hemolytic anemia is a popular subject in paleopathology literature. Its diagnosis was often linked to the presence of cribra orbitalia or porotic hyperostosis, which is defined as a "porotic or sieve-like appearance of bony orbital roofs" (Welker cited by Walper et al. 2004). The presence of cribra orbita-

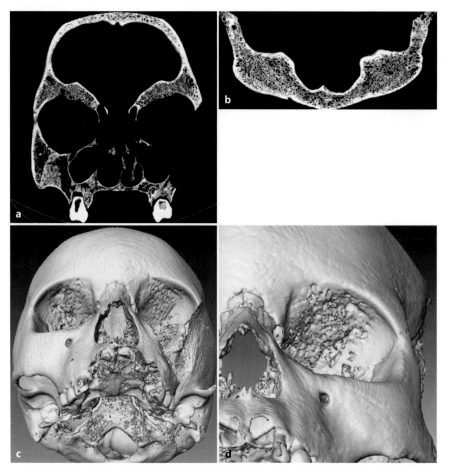

Fig. 4.76 a–d. Chronic anemia: Maya skull. **a** Coronal micro-CT. **b** Axial micro-CT. **c** 3D CT **d** 3D close-up on cribra orbitalia at the left orbit

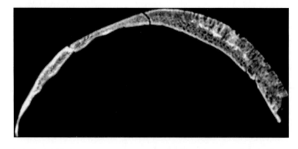

Fig. 4.77. Chronic anemia: Maya skull. Micro-CT of a skull fragment showing a "hair-on-end" pattern with thickening of the diploe and erosion of the outer table

References

Aegerter E (1975) Orthopedics Diseases, 4th edn. Saunders, Philadelphia

Aksoy M, Camli N, Erdem S (1966) Roentgenographic bone changes in chronic iron deficiency anemia. Blood 27:677–686

Aufderheide A, Rodriguez-Martin C (1998) The Cambridge Encyclopedia of Human Paleopathology. Cambridge University Press, Cambridge

Biesecker JL, Marcove RC, Huvos AG, Miké V (1970) Aneurysmal bone cysts. A clinical, pathological study of 66 cases. Cancer 26:615–625

Brothwell DR, Sandison AT (1967) Diseases in antiquity: a survey of the diseases, injuries, and surgery of early populations. Charles C. Thomas, Springfield

Brower AC, Flemmings DJ (1997) Arthritis in Black and White. Saunders, Philadelphia

Caffey J (1937) The skeletal changes in the chronic hemolytic anemias. Am J Roentgenol Radium Ther Nucl Med 37:293–324

Chapman S, Nakielny R (2003) Aids to Radiological Differential Diagnosis, 4th edn. Saunders, Edinburgh

Chhem RK (2006) Paleoradiology: imaging disease in mummies and ancient skeletons. Skeletal Radiol 35:803–804

Chhem RK, Ruhli FJ (2004) Paleoradiology: current status and future challenges. Can Assoc Radiol J 55:198–199

Chhem RK, Schmit P, Fauré C (2004) Did Ramesses II really have ankylosing spondylitis? A reappraisal. Can Assoc Radiol J 55:211–217

Cooney JP, Comdr EH, Crosby MC (1944) Absorptive bone changes in leprosy. Radiology 42:14

Crawford Adams J, Hamblen DL (1999) Outline of Fractures Including Joint Injuries, 11th edn. Churchill Livingstone, Edinburgh

Edeiken J, Freiberger RH, Jacobson HG, Martel W, Norman A, Resnick DL, Steinbach HL (1980) Bone Disease (Third Series) Syllabus. Professional Self-Evaluation and Continuing Program. American College of Radiology, Chicago

Esguerra-Gomez G, Acosta E (1948) Bone and joint lesions in leprosy: a radiologic study. Radiology 50:619–631

Exner S, Bogush G, Sokiranski R (2004) Cribra orbitalia visualized in computed tomography. Ann Anat 186:169–172

Faget GH, Mayoral A (1944) Bone changes in leprosy: a clinical and roentgenologic study of 505 cases. Radiology 42:1–13

Forrester M, Brown JC (1987) The Radiology of Joint Disease. Saunders, Philadelphia

Freiberger RH, Edeiken J, Jacobson HG, Norman A (1976) Bone Disease (Second Series) Professional Self-Evaluation and Continuing Program. American College of Radiology, Chicago

Galloway A (1999) Broken bones: anthropological analysis of blunt force trauma. Charles C. Thomas, Springfield

Greenblatt C, Spiegelman M (eds) (2003) Emerging Pathogens. Oxford University Press, Oxford

Griffith HJ (1987) Basic Bone Radiology, 2nd edn. Appleton Lange, Norwalk

Haas CJ, Zink A, Palfi G, Szeimies U, Nerlich AG (2000a) Detection of leprosy in ancient human skeletal remains by molecular identification of Mycobacterium leprae. Am J Clin Pathol 114:428–436

Haas CJ, Zink A, Molnar E, Szeimies U, Reischl U, Marcsik A, Ardagna Y, Dutour O, Pálfi G, Nerlich AG (2000b) Molecular evidence for different stages of tuberculosis in ancient bone samples from Hungary. Am J Phys Anthropol 113:293–304

Harris HA (1930) Lines of arrested growth in the long bones in childhood: the correlation of histological and radiographic appearances in clinical and experimental conditions. Br J Radiol 18:622–640

Helms AC (2005) Fundamentals of Skeletal Radiology, 3rd Edn. Philadelphia, PA

Herman B, Hummel S (2003) Ancient DNA can identify disease elements. In: Greenblatt C, Spiegelman M (eds) Emerging Pathogens. Oxford University Press, Oxford, pp 143–150

Huvos AG (1991) Bone Tumors. Diagnosis, Treatment, and Prognosis, 2nd edn. Saunders, Philadelphia

Keats TE (1988) An Atlas of Normal Developmental Roentgen Anatomy. Year Book Medical Publishers, Chicago

Lachman RS, Hooshang Taybi H (1996) Radiology of Syndromes, Metabolic Disorders, and Skeletal Dysplasias, 4th edn. Mosby, St. Louis

Laredo JD, Reizine D, Bard M, Merland JJ (1986) Vertebral Hemangiomas: Radiologic Evaluation. Radiology 161:183–189

Lichtenstein L (1972) Bone Tumors. Mosby, Saint Louis

Lovell NC (1997) Trauma in Paleopathology. Yearb Phys Anthropol 40:139–170

Madewell JE, Ragsdale BD, Sweet DE (1981) Radiologic and pathologic analysis of solitary bone lesions: part II – internal margins. Radiol Clin North Am 19:715–747

Manchester K (2002) Infection bone changes of leprosy. In: Roberts CA, Lewis ME, Manchester K (eds) The Past and Present of Leprosy: Archaeological, Historical, Paleopathological and Clinical Approaches. British Archaeological Reports International Series 1054, Archaeopress, Oxford pp 69–72

Milgram JW (1990) Radiologic and Histologic Pathology of Non Tumorous Diseases of Bones and Joints. Vol 1 and Vol 2. Northbrook Publishing, Northbrook

Moller-Christensen V (1961) Bone Changes in Leprosy. Munksgaard, Copenhagen

Ortner D J (2003) Identification of Pathological Conditions in Human Skeletal Remains, 2nd edn. Academic Press, London

Pastushyn, AI, Slin'ko E, Mirzoyeva GV (1998) Vertebral Hemangiomas: Diagnosis, management, natural history and clinicopathological correlates in 86 patients. Surg Neurol 50:535–547

Powel ML, Cook DC (2005) The Myth of Syphilis: The Natural History of Treponematosis in North America. University of Florida Press, Gainesville

Ragsdale BD (1993) Morphologic analysis of skeletal lesions: correlative of imaging studies and pathologic findings. Adv Pathol Lab Med 6:445–490

Ragsdale BD, Madewell JE, Sweet DE (1981) Radiologic and pathologic analysis of solitary bone lesions. Part II – periosteal reactions. Radiol Clin North Am 19:749–783

Reimann F, Kayhan V, Talati U, Grokmen (1975) X-ray and clinical study of the nose, sinuses and maxilla in patients with severe iron deficiency diseases. Laryngol Rhinol Otol 54:880–890

Resnick D, Kransdorf MJ (2005) Bone and Joint Imaging. Elsevier Saunders, Philadelphia

Roberts C, Manchester K (2005) The Archaeology of Disease, 3rd edn. Cornell University Press, Ithaca

Rogers J, Waldron T (1989) Infections in paleopathology: the basis of classification according to most probable cause. J Archaeol Sci 16:611–625

Rogers J, Waldron T (1995) A field guide to joint disease in archaeology. John Wiley, Chichester

Rogers J, Waldron T, Dieppe P, Watt I (1987) Arthropathies in paleopathology: the basis of classification according to most probable cause. J Archaeol Sci 14:179–193

Ruhli FJ, Chhem RK, Boni T (2004) Diagnostic paleoradiology of mummified tissue: interpretation and pitfalls. Can Assoc Radiol J 55:218–227

Schmorl G, Junghans H (1956) Clinique Radiologie de la Colonne Vertebrale Normale et Pathologique. Doin Cie, Paris

Schmorl G, Junghans H (1971) The Human Spine in Health and Disease, 2nd edn. Grune and Stratton, New York

Schultz RJ (1991) The Language of Fractures, 2nd edn. Williams and Wilkins, Baltimore

Schultz M, Roberts CA (2002) Diagnosis of leprosy in skeletons from an English later medieval hospital using histological analysis. In: Roberts CA, Lewis ME, Manchester K (eds) The Past and Present of Leprosy: Archaeological, Historical, Paleopathological and Clinical Approaches. British Archaeological Reports International Series 1054. Archaeopress, Oxford, pp 89–104

Sebes JI, Diggs LW (1979) Radiographic changes of the skull in sickle cell anemia. AJR Am J Roentgenol 132:373–377

Steinbock RT (1976a) Paleopathological Diagnosis and Interpretation: Bone Diseases in Ancient Human Populations. Charles C. Thomas, Springfield

Steinbock RT (1976b) Hematologic disorders – the anemias. In: Steinbock RT (ed) Paleopathological Diagnosis and Interpretation: Bone Diseases in Ancient Human Populations, 1st edn. Charles C. Thomas, Springfield, pp 213–248

Stuart-Macadam P (1992) Porotic hyperstosis: a new perspective. Am J Phys Anthropol 87:39–47

Sweet DE, Madewell JE, Ragsdale BD (1981) Radiologic and pathologic analysis of solitary bone lesions: part III – matrix patterns. Radiol Clin North Am 19:785–814

Walper U, Crubezy E, Schultz M (2004) Is cribra orbitalia with anemia? Analysis and interpretation of cranial pathology in Sudan. Am J Phys Anthropol 123:333–339

Zink A, Haas CJ, Reischl U, Szeimies U, Nerlich AG (2001) Molecular analysis of skeletal tuberculosis in an ancient Egyptian population. J Med Microbiol 50:355–366

Chapter 5

Paleoradiology in the Service of Zoopaleopathology

5

Don R. Brothwell

Prehistorians and medical historians have been interested in finding and describing evidence of diseases in human remains since the 19th century, but in contrast animal remains have been neglected. Admittedly, ancient nonhuman pathology was brought into reviews by Moodie (1923) and Pales (1930), and a larger study appeared by 1960 (Tasnádi 1960). Alas, the latter work by Tasnádi (1960), "Az Osállatok Pathologiája," appeared only in Hungarian and has been largely ignored. So it was not until 1980 that a general introductory work appeared on "Animal Diseases" (Baker and Brothwell 1980), which is slowly being updated by other findings of pathology and new discussions (Brothwell 1995; Luff and Brothwell 1993).

Clearly, we now need more evidence of zoopaleopathology, both generally in actual pathological remains, and by applying veterinary and molecular techniques to their investigation. X-rays provide one important line of investigation, producing conventional radiographs or computed tomography (CT) scans, although the latter are only slowly being applied to zooarcheological material. Recently, for instance, a series of Egyptian animal mummies were CT scanned, to reveal basic details on the animals and possible evidence of abnormality. Similarly, pathological bones of mammoth from an Upper Pleistocene site at Lynford in East Anglia (UK), were scanned to reveal more about the pathology and to check for possible fragments of stone tools being used in hunting by contemporary Neanderthalers (Brothwell et al. 2006) (Figs. 5.1 and 5.2). In this chapter, an attempt is made to review, albeit briefly, the range of pathology, especially of bones and teeth, which can be revealed or confirmed by radiographic study. It is clearly important to base diagnoses on the same criteria employed by veterinary radiologists on living species, but at the same time to be aware that archeological material brings with it special diagnostic problems. In particular, broken and incomplete material is usual and postmortem changes can occur. Nevertheless, it is hoped the following will show that differential x-ray-supported diagnoses are possible.

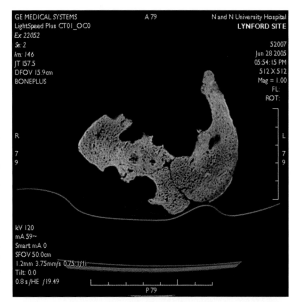

Fig. 5.1. Computed tomography (CT) scan section of rib pathology in a mammoth from Lynford, England

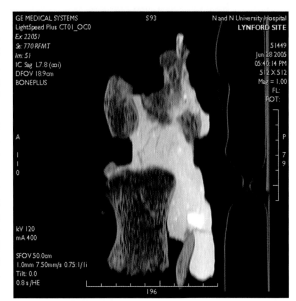

Fig. 5.2. CT scan detail of pathology in the caudal vertebral area of a mammoth

5.1
Congenital Abnormalities

Inborn defects in skeletal development can vary considerably from total skeletal modification to very minor changes to one or a few bones. The reason for the error in bone growth may be genetic, of unknown etiology, or of intrauterine origin, including infections such as bovine viral diarrhea, which can stunt growth. In archeological material, unless the whole skeleton is preserved, diagnosis is more likely to have to be on incomplete material and even damaged skeletal elements. In such situations, radiological detail can often assist in understanding the nature of the defect, and in revealing its internal detail.

While congenital abnormalities (Johnson and Watson 2000) in earlier domestic or wild species have an interest in terms of the general biology of these species, any noticeably recurring anomalies could potentially be indicating inbreeding. In domestic species, such evidence gives information on the nature of animal practices in such communities. The abnormalities can in these cases take the form of rib shortening, sternal variation, hydrocephaly, jaw reduction, and vertebral malformation (Chai 1970). In terms of the general occurrence of skeletal defects, not all oc-

cur to the same degree in domestic groups (Table 5.1). Those listed are only a select list, and other species may display other skeletal defects. Angular limb deformities may be prevalent in llamas, as are polydactyly, syndactyly, and malformations of the mandible (Fowler 1989).

Some congenital defects will be much more obvious than others, even when only a single bone is present. When the element is fragmentary, radiographic details may help to clarify the anomaly and confirm the diagnosis. Dwarfism in archeological cattle, when extreme, is difficult to identify, and x-rays can assist in confirming the bone and anomaly (Fig. 5.3). Polydactyly or monodactyly would be easy to identify provided the distal cannon joint was intact (Leipold 1997). It is important to remember that one male who carries a dominant mutant gene that influences skeletal defects may cause significant disease in a population (both now and in the past). An example of this is seen in the occurrence of osteogenesis imperfecta (Holmes et al. 1964), where one ram with the gene affected a series of lambs in the flock. Radiographically, the long bones of the young animals displayed abnormal structure, with generally thin, poorly mineralized cortical bone. The skull was similarly affected, with the central areas of the frontal and parietal bones almost completely radiolucent.

Table 5.1. Selected skeletal conditions of congenital origin in domestic species (modified from Fowler 1989). Where the cell is left blank, there was no information available. *Y* inheritance confirmed, *S* inheritance suspected, *U* etiology unknown

Condition	Bovine	Equine	Ovine	Caprine	Porcine	Canine	Feline
Ankylosis (carpus)		U					
Arthrogryposis	Y	Y	Y		Y		
Femur shortened							Y
Hemivertebrae	S				Y	U	
Metacarpal shortening						U	Y
Polydactyly	Y	U	Y	Y	S	Y	Y
Scoliosis	U		U				
Syndactyly	Y		U		U	Y	
Encephalomeningocele			U				U
Hydrocephalus		U	U		S	Y	Y
Brachygnathia (mandible)	Y	Y	Y	Y	U	Y	
Micrognathia			Y		U		
Brachygnathia (maxilla)					U	Y	
Cleft palate	Y	U	Y		U		Y
Dwarfism	Y	Y	Y	Y	Y	U	

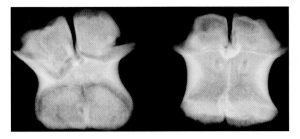

Fig. 5.3. X-rays of longitudinally reduced long bones of a "bull-dog" calf

5.2
Summary of Radiological Features of Congenital Abnormality

The following summary outlines the range of congenital variation in the skeleton, some of which has already been noted in archeological cases, and others that could be identified in the future (albeit in fragmentary remains).

5.2.1
The Skull

5.2.1.1
Encephalomeningocele

Encephalomeninogocele is seen as a defective cranial formation, medially(Fig. 5.4). The rounded cranial defect may be represented archeologically by only part of the defective skull. In x-ray, the margins of the defect can be seen to be rounded normal bone.

5.2.1.2
Hydrocephalus

Enlargement of the brain box (Fig. 5.5) with much expanded (and possibly thinned) frontal, parietals, and occipital plates. As fragments, change in the enlarged and remodeled bones might still be demonstrated.

5.2.1.3
Brachygnathia and Micrognathia

This is retarded development of the jaws, singly or upper and lower together. In fragments, external and internal detail may indicate smaller bone size.

5.2.1.4
Cleft Palate

The incomplete formation of the palate in the midline (Fig. 5.6). The margins of the defect should appear

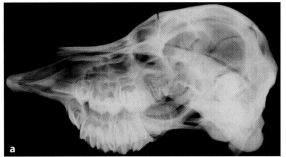
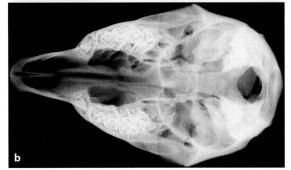

Fig. 5.4 a,b. Medial opening in the upper posterior aspect of the cranial vault as a result of **a** meningocele

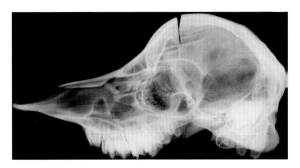

Fig. 5.5. Radiographic detail of a calf with hydrocephaly, showing an enlarged brain area

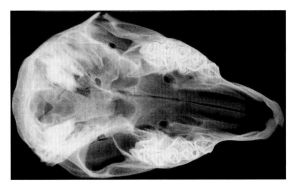

Fig. 5.6. Calf with complete absence of palate (no bone between the teeth)

radiographically as smooth normal bone. Survival chances are not good with this condition.

5.2.1.5
Cerebral Hernia

This condition, which has been reported in a Roman chicken skull (Brothwell 1979), is distinctive, especially in lateral x-ray. The endocranial cavity is clearly enlarged in medial contour, with the anterior area expanded and even perforating the external surface (Fig. 5.7).

5.2.2
The Postcranial Skeleton

5.2.2.1
Dwarfism

Various forms are characterized by growth reduction, especially of longitudinal growth in the long bones. Dyschondroplasia in a humerus is known from a case at the Knap of Howar in Scotland (Baker and Brothwell 1980). The bulldog calf anomaly results in a more extreme reduction in growth.

5.2.2.2
Hip Dysplasia

Major femur head and pelvic acetabulum defects are seen as a result of hip joint maldevelopment. Ultimately, the femur head may ride out of the acetabulum and form a "secondary joint" on the ilium. The acetabulum may be poorly formed and shallow, and following femoral dislocation can show secondary remodeling (Fig. 5.8). At least two cases are described archeologically. It is most likely to occur in dog remains (Murphy 2005).

5.2.2.3
Hemivertebrae

Abnormal wedged "half-vertebrae" are one form of congenital abnormality, and can cause scoliosis (anomalous side-to-side curvature) of the spine. Fragmentary vertebrae, especially if also fused, may not clearly show the condition, and x-ray may assist in diagnosis. A case has been described from Iron Age Danebury in England (Brothwell 1995).

5.2.2.4
Arthrogryposis

A congenital condition causing persistent flexure and contracture of a joint. Malformation of the bones of

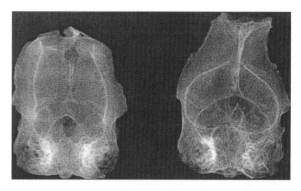

Fig. 5.7. Roman chicken skulls in dorsoventral view, the one on the left displaying an expanded endocranial (brain) area, indicating cerebral hernia. X-ray courtesy of T. and S. O'Connor

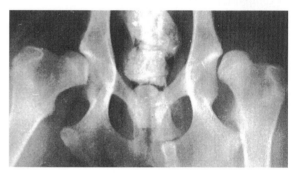

Fig. 5.8. Severe hip dysplasia in a dog. Note the poorly formed acetabulum

the joint occurs. Trauma must be distinguished, as well as other arthropathies.

5.2.2.5
Syndactyly

This is known to occur in various species and affects mainly the distal digits of the foot, which can be totally fused (Fig. 5.9). In eroded and incomplete archeological cases, radiographic detail may help to confirm the diagnosis. In the case of a pig example from a British site (Osbourne House), the distal phalanges were well fused. Other bones of the distal limbs can also be fused, as in, for instance, the third and fourth carpals of an ancient pig from Garton Slack in Yorkshire (Baker and Brothwell 1980).

5.2.2.6
Other Conditions

There are several other congenital defects that have been noted in archeological material (Figs. 5.10–5.12), including bifid ribs, enlarged foramina (in various bones), and nontraumatic clefts in joints. However,

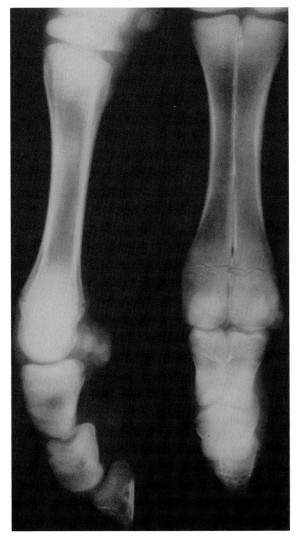

Fig. 5.9. Syndactyly in a calf. Modified from Bargai et al (1989)

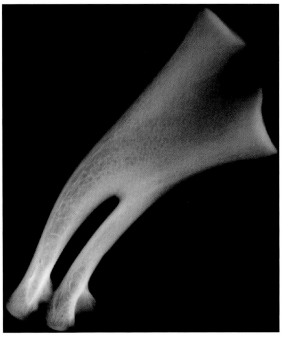

Fig. 5.10. Conjoined ribs in a young calf (congenital)

the previous descriptions at least provide a survey of the degree of variation, and the fact that x-rays often assist in making a firm diagnosis. Clearly, there is a need to understand the degree of normal variation in order to identify a congenital abnormality.

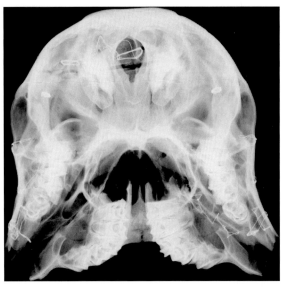

Fig. 5.11. Internal detail of a two-headed calf, dorsoventral view

5.2.3
Nutritional and Metabolic Conditions

Archeologists have a long-term interest in the environment, changes through time, and its varying impact on the growth and the well being of earlier human populations, and their associated domesticates. Besides the minor influences of diet, temperature, and altitude on growth, where animals are subject to more ecological extremes, skeletal changes may extend into the pathological, with several well-defined conditions resulting when the environmental stresses continue for long enough. In all cases, radiographic study may assist in diagnosis, especially if the material is incomplete.

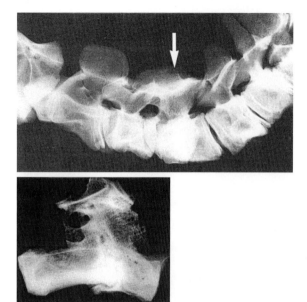

Fig. 5.12. Congenital deformity in vertebrae; comparing a recent case (top) with a Iron Age case (bottom)

Fig. 5.13. Medieval sheep femur with osteoporotic bone loss and thin cortical bone of the shaft

5.2.3.1
Osteoporosis (Osteopenia)

Osteoporosis (Fig. 5.13) is a condition that may result from several more specific diseases such as hyperparathyroidism, starvation, parasitism, hypothyroidism, and anemia (Dennis 1989). The result of this is that the bone is normal but deficient in amount. There may be cortical thinning and a reduction in bone density. The medullary cavities will also appear wider than normal if the animal has been osteoporotic for some time before death. Harris (growth arrest) lines may confirm long-term environmental stress. Orbital cribra, in species other than humans, appear to be rare, although it could possibly be associated with renal failure and its consequences in cats. Limited regional osteoporosis may occur if, for instance, there is long-term disuse of a limb after trauma. In cattle, osteoporosis usually develops in growing calves, and some of the fractures seen may in fact be a sequel to

osteoporosis (Greenough et al. 1972). Senile osteoporosis can occur in old animals, but seems unlikely to be a common cause, as in the past, few would have been allowed to reach senility. In young animals, osteoporosis may also be associated with rickets, and can result from placing newly weaned animals on poor pasture.

5.2.3.2
Rickets

This skeletally damaging condition is seen in young animals and results from a deficiency of dietary phosphorus or vitamin D. While uncommon in farming communities today, it may not have been in the past. Seen in x-ray, the growth plates of the young animal are wider than normal, so that the diaphyseal ends appear to be splayed out. There is also a ragged margination and epiphyseal irregularity. In the later stages, the poorly mineralized bone allows long bones to bend in weight bearing. Rickets can occur at any age during growth. The equivalent in the adult animal is osteomalacia, with decalcification of the bone matrix. It is likely to occur particularly in pregnant and milking cows, and particularly in areas of poor soil conditions. The radiographic evidence is the same as in rickets. The extreme form of nutritional disease, beyond osteomalacia, is seen in secondary hyperparathyroidism, where cranial thickening may also occur (du Boulay and Crawford 1968).

5.2.3.3
Hypervitaminosis A

This has been well described in cats (Baker and Hughes 1968), but is known to at least occur in dogs and humans (Fig. 5.14). It is particularly the result of animals being fed on a diet that is rich in liver. The excess vitamin A stimulates the formation of subperiosteal new bone, and can be massively developed on the limb and vertebral joints. Some bone remodeling could occur as a result of limb immobility, but has not been described. A differential diagnosis would need to exclude in particular massive osteoarthrosis of the limbs and spondylosis of the spine.

5.2.3.4
Hypothyroidism

While there is a congenital form described in horses, dogs, and cats, it is rare, and the usual form is due to iodine shortage in the environment. Only in the severe form of cretinism could this be recognized archeologically, and it might depend on the amount and region of the skeleton remaining. This thyroxine defi-

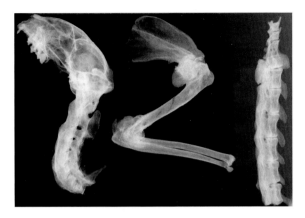

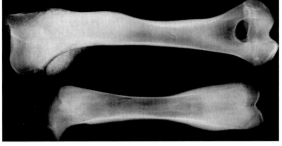

Fig. 5.14. Extensive union of bones at joints in a cat, due to vitamin A poisoning

Fig. 5.15. Evidence of scurvy in a dog (uncommon). The proximal humerus shows an early stage ossifying hematoma. The distal femur displays a smooth swelling, indicating a much older bleeding and ossification

ciency results in dwarfism, with delayed maturation, short and broad long bones with thick cortices, vertebral changes, and kyphosis. However, the animals would probably die very young.

5.2.3.5
Juvenile Scurvy (Hypertrophic Osteodystrophy)

Currently, this condition is only seen in large breeds of dogs, but could have a very different history in certain species. Long bones are especially affected in the young animals (Fig. 5.15). The cause appears to be uncertain, although lack of vitamin C is one factor. Radiographically, the ends of the diaphyses may display radiolucent areas and regions of increased opacity. Subperiosteal hemorrhage may occur, followed by the ossification of one or more ossifying hematomas. In primates and guinea pigs, true scurvy can occur, with both skeletal and dental pathology resulting.

5.2.3.6
Osteodystrophia Fibrosa

This is another nutritional deficiency condition, and can follow osteoporosis. It is well noted in ruminants, but also appears in horses, carnivores, and pigs (Andrews 1985). In the past it was known as "big head," especially in horses, where it was associated with bran feeding. While its pathogenesis appears to be complex, the calcium/phosphorus balance is critical. It can be seen as an extension of secondary hyperparathyroidism, and appears to be most rapidly developing in goats. A progressive swelling occurs in the mandible, and to a lesser extent the maxillae, nasals, zygomatic bones, and frontal area. There may also be lingual angulation of the cheek teeth, which may also become loose.

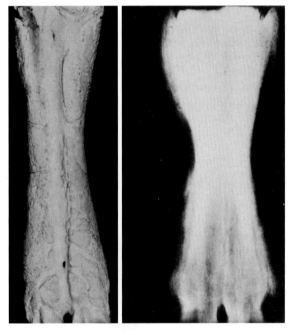

Fig. 5.16. Cortical bone changes resulting from chronic fluorosis. Cow metatarsals, showing dense bone in x-ray (right). After Jubb et al. (1985)

5.2.3.7
Fluorosis

While this is seen in industrial society as a pollution problem, in fact fluorine in the environment can be high naturally. In young animals, fluorine may be combined with skeletal mineral in all areas, but later may be concentrated. Subperiosteal hyperostoses can be bilateral and extensive. In dry-bone pathology, the bone reaction looks surprisingly like hypertrophic pulmonary osteopathy (Fig. 5.16). Articular surfaces are normal, although mineralization may occur in tendons and ligaments near joints (Jubb et al. 1985).

During the development of the teeth, mottled areas of enamel may form, and in severe degrees of fluorosis this tissue can be chalky and opaque, leading to anomalous wear. Radiologically, the subperiosteal new bone appears to be dense and opaque, with a coarsened trabecular pattern.

5.2.3.8
Harris Lines

While not a distinct pathological entity, mention should be made here of radio-opaque transverse trabeculae that have been noted in the bones of various mammals, particularly humans. They have become known as Harris lines because they were first discussed by the anatomist H.A. Harris (1931), and can be seen in some x-rays as lines at right angles to the long axis of long bones. They also occur in other bones, and in the pig have been noted where there is considerable growth and remodeling at the angle of the mandibular ramus. Their etiology is debatable, but could be associated with nutritional stress and infection during growth (Platt and Stewart 1962).

Other examples of environmental stress are given in Figs. 5.17–5.19

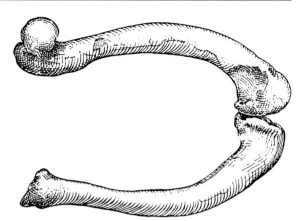

Fig. 5.18. Femur and tibia from an ancient Egyptian baboon, probably indicating osteomalacia. Such cases deserve radiographic evaluation. After Moodie (1923)

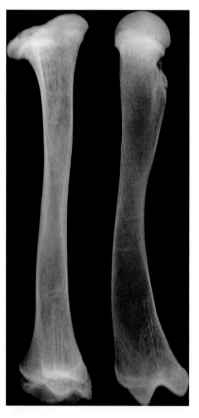

Fig. 5.19. Femur and tibia of a young dog, with poor bone formation caused by secondary (nutritional) hyperparathyroidism

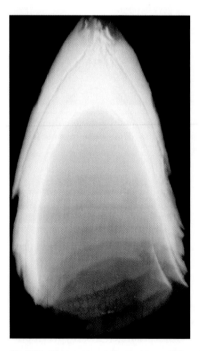

Fig. 5.17. Environmental stress resulting in defective horn growth in a Roman calf from Vindolanda, England. Underlying bone removed

5.3
Trauma

Injuries to the skeleton, usually showing quite advanced states of repair, are relatively common in bird and mammal remains. However, it is difficult

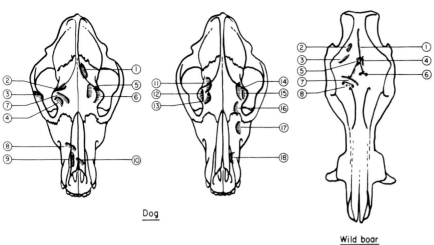

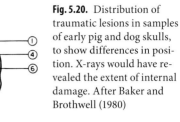

Dog

Wild boar

Fig. 5.20. Distribution of traumatic lesions in samples of early pig and dog skulls, to show differences in position. X-rays would have revealed the extent of internal damage. After Baker and Brothwell (1980)

in zooarcheological material to detect perimortal fractures; indeed it is suspected that fractured bones displaying very early stages of healing are also all too often missed. The situation is not made any easier by the fact that such trauma evidence is often incomplete, and may be butchered food residues. In the case of possible well-healed fractures, it is first of all important to establish by x-ray that the swelling is traumatic and not osteomyelitic or neoplastic. Radiography will also establish the degree of overlap in the fractured bone, or the angular relationships between the broken pieces. In the case of vertebral bodies, wedging may be due to compressional collapse, or alternatively infection leading to vertebral collapse. In various respects then, radiographic evaluation is essential to the full understanding of the trauma.

Unlike trauma in modern veterinary situations, ancient examples of injury do not include those caused by cars, racing, or gunshot (except in postmedieval times). Nevertheless, in past societies, horse kicks, male dominance fighting, simple accidents, and human hunting and farm brutality can all result in skeletal trauma. The big question is to what extent can the various causes be identified? The anatomical position of the fracture, the extent of the breakage and displacement, the degree of healing, even the inclusion of metallic or other fragments, may all provide clues useful to a reconstruction of the traumatic event. An example of the differences that can be shown to occur is seen in samples of cranial trauma in wild boar and domestic dogs (Fig. 5.20), where the former display brain-box injuries, and the dogs, orbital and snout lesions (Baker and Brothwell 1980).

A good review of fracture types and healing, as revealed in x-ray, is provided by Robert Toal and Sally Mitchell (Toal and Mitchell 2002). In zooarcheology, it is not usual to provide detailed descriptions of trauma, but supported by x-rays, this is important

to do. The range of fractures may be summarized as follows:

1. Incomplete, "greenstick," fissure or fatigue fractures. As the terms imply, these may show a partial break with bending, a longitudinal fissuring without displacement, or bone-stressed microfracturing. These are not easy to identify.
2. Complete simple fractures show a break fully through the bone. The fracture line may be transverse, oblique or spiral.
3. Comminuted fractures display multiple lines, extending from one plane.
4. Segmented fractures show more dispersed fragmentation.
5. Impacted, or crushing, fractures are most usual in vertebrae.
6. Depressed fractures are seen where there is collapse of bone in the skull.
7. Avulsion fractures are usually linked to traction stress, with fragmentation usually.
8. Chip fractures: these are probably not of importance zooarcheologically.

5.3.1
Fracture Healing

The quality of healing and the time taken to heal are factors of interest to zooarcheologists for two reasons. First, if the fracture is well healed and in good alignment, in some situations it might suggest good husbandry in farming communities, and care for the traumatized animal. Care is needed in interpretation here, as Adolf Schultz (Schultz 1939) described some surprisingly well-healed fractures in a series of wild primates. Second, the time taken to produce the degree of healing in a relatively "new" fracture, is suggestive of the amount of special care the animal received, or the time the animal was allowed to live before eventually being put down. The pathology found at Iron Age Danebury (Brothwell 1995) suggested strongly the intentional care of animals that might have been better dispatched. It should be noted that healing times of fractures can only be sensibly estimated if there is no associated pathology, such as secondary nutritional hyperparathyroidism, or at the site of a destructive neoplasm. Some degree of movement at the fracture may also deter healing, although a callus may form but not unite the bone ends. As a rough guide to healing, it should be noted that within 10 days, some bone end rounding and demineralization will have occurred. Within 1 month, a mature callus will be clearly forming around the fracture ends and sides. After 3 months, the fracture should be united, with callus surface remodeling. Times will be extended to some degree in old animals.

In Figs. 5.21–5.27, a range of archeological fracture cases are shown, showing various degrees of healing distributed over the skeleton. Some, such as fractures to the pelvis and mandibular ramus are surprising, and raise the question of the quality of husbandry in these early societies. There is probably more literature on zooarcheological fractures than on any other aspect of ancient animal health. In the case of a fossil hyena described by Vlcek and Benes (1974), x-rays provided further internal information on the nature of the trauma and the area behind the eyes and extending along the region of the sagittal crest. Deformities to the temporal muscles had influenced skull morphology and resulted in eventual asymmetry (Fig. 5.28). The value of understanding vertebral trauma as revealed by x-rays in modern horses enables a better understanding of trauma in ancient horses (Jeffcott and Whitwell 1976) (Fig. 5.29). Udrescu and van Neer (2005) have raised the interesting question of the possible intervention of human groups in the treatment of domestic species in the past. They review a range of fractures in domestic and wild species, employing radiography to some extent, and conclude that there is as yet no good evidence to support this hypothesis.

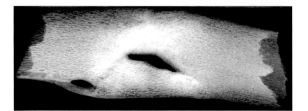

Fig. 5.23. Cow rib fragment, with healed fracture

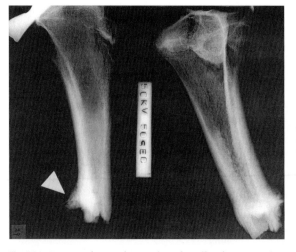

Fig. 5.21. Fractured cow tibia, with early callus formation (arrowhead). Iron Age

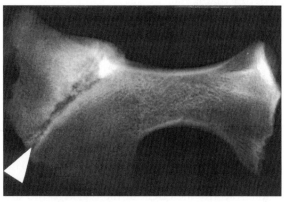

Fig. 5.24. Young sheep with a fractured but healing pelvis (iliac blade; arrowhead), Danebury

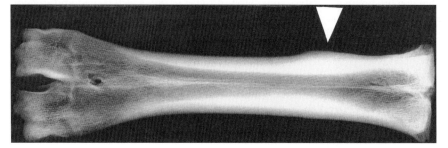

Fig. 5.22. Cow cannon bone with restricted cortical expansion (arrowhead), possibly indicating an ossifying hematoma. Dragonby, UK

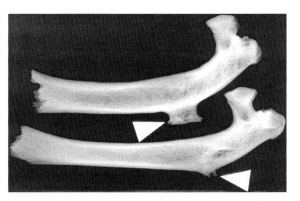

Fig. 5.25. Horse ribs displaying minor trauma and soft tissue ossification (arrowheads)

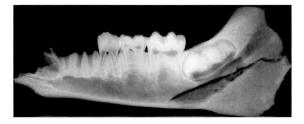

Fig. 5.26. Young pig mandible, with healing fracture. Danebury Iron Age

5.3.2
Infection

Osteitis, periostitis, and osteomyelitis are caused by the deposition and growth of microorganisms in bone. The three terms indicate the extent of the infection, from a restricted subperiosteal reaction to the total involvement of bone. The microbes arrive as the result of a direct skeletal injury, by infection of neighboring tissues, or via the circulation of the blood. The infection may be very restricted or widespread in the skeleton. Radiographic evaluation is critical to a careful differential diagnosis, and in particular, neoplastic changes need to be excluded. Septic arthritis, the involvement of joints, has been considered with other arthropathies.

In zooarcheological material, periostitis is seen more commonly than more extensive osteomyelitis. It probably affects immature bovids more than other species (Bargai et al. 1989). Examples of the variable radiographic changes that can be seen in zooarcheological material are shown in Figs. 5.31–5.38. The following are more specific conditions that can occur, and where radiography is of diagnostic value.

Fig. 5.27. Medieval dog ulna and radius, displaying fracture and healing

Fig. 5.28. Trauma to a fossil hyena skull. (After Vlcek and Benes, 1974.)

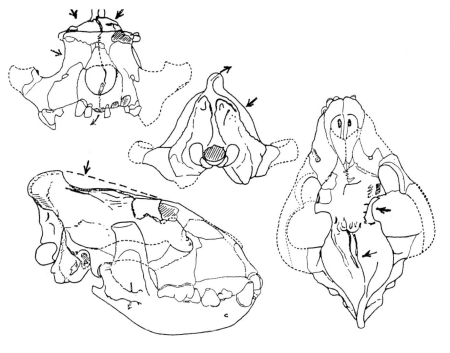

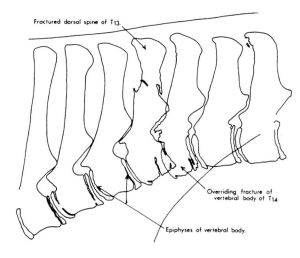

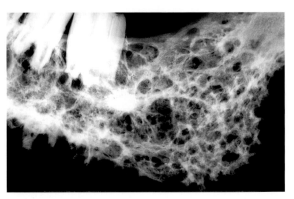

Fig. 5.31. Severe bone changes caused by actinomycosis. Modern cow mandible

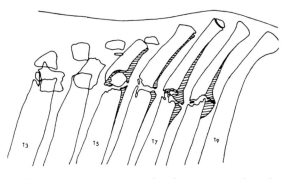

Fig. 5.29. Vertebral trauma in modern horses, as a guide to the interpretation of ancient examples. a Crush fractures involving T13+14 in a young filly. b Multiple fractures from a fall in a young gelding. After Jeffcott and Whitwell (1996)

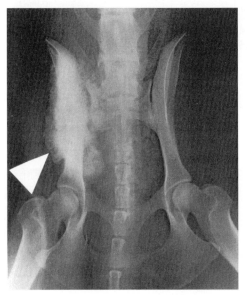

Fig. 5.32. Recent pelvis displaying changes caused by coccidioidomycosis (arrowhead)

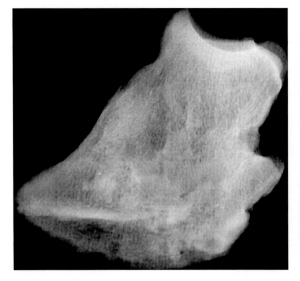

Fig. 5.30. Phalangeal union caused by long term infection ("foul-in-the-foot")

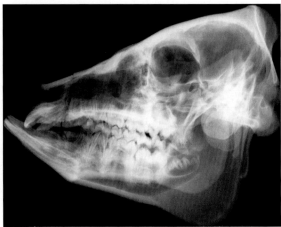

Fig. 5.33. Retarded snout growth in a young pig, indicating atrophic rhinitis

5.3.2.1
Interdigital Necrobacillosis

Also known as "foul-in-the-foot," this is initially a soft-tissue infection that can be caused by several microbes. It can have a high incidence in dairy cattle, affecting various ages. Predisposing factors include standing in dung and mud, yards with sharp stones or stubble, and possibly hereditary factors. Lesions to the interdigital space allow microbes to enter the foot. Eventually the distal bones of the foot may become involved (Fig. 5.30), and there can be union of bones and a general appearance of osteomyelitis.

5.3.2.2
Vertebral Osteomyelitis

It should be mentioned here that whereas in human groups, a destructive osteomyelitis of the vertebrae is most commonly associated with tuberculosis. In young pigs, cattle, and lambs, it is more likely to be another microbe, including *Brucella suis* and *Corynbacterium pyogenes*. As yet (May 2005), there is little discussion in the archeological literature.

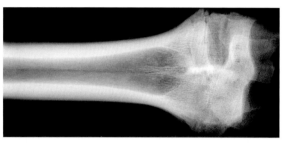

Fig. 5.36. Horse from Saxon Southampton, first considered to display spavin, but more likely to have a joint infection

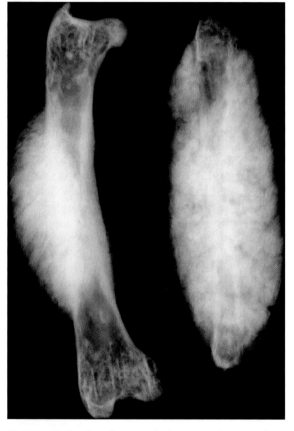

Fig. 5.34. Chicken long bones with dense extra bone resulting from osteopetrosis

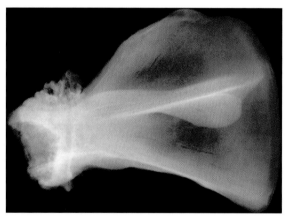

Fig. 5.37. Pig scapula showing marked septic arthritis at the glenoid fossa

Fig. 5.35. Young dog with forelimb osteomyelitis and sequestrum (arrowhead). After Morgan (1988)

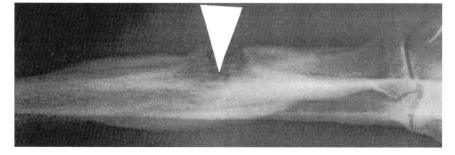

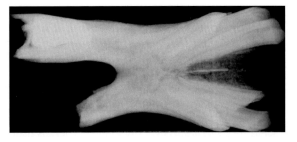

Fig. 5.38. Osteomyelitis of the left mandibular body of a Danebury horse

5.2.3.3
Actinomycosis

Although it is claimed to have been noted in ancient cattle jaws, no certain cases of actinomycosis, also known as "lumpy jaw," have yet been identified (Jensen and Mackey 1979). This noncontagious osteomyelitis produces a characteristic granulomatous mass of new bone (Fig. 5.31), usually in the mandible. Today, there appears to be no breed, gender, or age bias. Radiographically, the swollen bone mass appears to be filled with well defined "soap bubbles" (multiple osteolytic foci).

5.2.3.4
Coccidioidomycosis

Unlike the previous mycotic infection, the new bone in this condition is relatively dense and extends from the cortical surface. In x-ray, it can appear somewhat spicular in form (Maddy 1958; Morgan 1988), and a differential diagnosis must take into account neoplasms and other forms of infection (Fig. 5.32). It appears to have a predilection for dogs. Geographically, animals of the southern United States and Mexico are most likely to be affected (Fink 1985). The microbe appears to prefer dogs and humans, but is noted also in pigs, sheep, horses, cats, and rodents.

5.2.3.5
Atrophic Rhinitis

This is a well-known condition in pigs, but can occur in other species, especially cats and dogs, but in these other species it is a far less severe disorder. The most common lesion is the atrophy of the turbinates in the nose, not easily recognizable in archeological material. There may be an associated osteoarthrosis of the temporomandibular joint. Most noticeable in a percentage of the cases is a retarded snout (Fig. 5.33), sometimes with lateral twisting and distortion. X-rays of the whole skull, or even just the snout, may reveal inner anomalies and the relative shortening of

the snout. Genetic factors may influence the degree of development of the disease (Kennedy and Moxley 1958; Penny and Mullen 1975).

5.2.3.6
Osteopetrosis

Numerous archeological chicken bones display pathology the result of infection by avian leucosis viruses (Fig. 5.34). The disorder mainly affects the long bones, and is the result of excessive osteoblast proliferation. In x-rays, there is clearly gross thickening, with the dense cortical bone extending inwards into the medullary cavity, as well as outwards along much of the shaft (Payne 1990). The virus appears to have arrived in Britain during Roman times, and spread geographically in chicken populations by the medieval period (Brothwell 2002).

5.4
The Arthropathies

In zooarcheology, there is rarely the possibility of viewing joints intact. Thus it is not possible to evaluate, for instance, the degree of narrowing of the joint space or the positioning of one bone against another within the joint. On the other hand, the pathology of the dry bone may be very clear, and radiographic detail can add significantly to the joint interpretation. Classification of joint disease is not completely the same for humans and other species, but there is considerable overlap (Pedersen et al. 2000). Joint changes linked to gout, as well as diffuse idiopathic skeletal hyperostosis, appear to be distinctly human conditions (Rogers et al. 1987). Overall, the radiographic changes indicative of some form of arthropathy in ancient material, would particularly include (Allan 2002):

1. Anomalous bone in the joint area.
2. Decreased or increased subchondral bone opacity.
3. Subchondral bone cyst formation.
4. Altered perichondral bone opacity.
5. Perichondral bone proliferation.
6. Mineralization of joint soft tissues.
7. Intra-articular calcified bodies.
8. Joint malformation.

The extent to which these signs are represented depends on the nature of the subchondral bone loss, which may be smooth and restricted or may be ragged and irregular and result in considerable bone destruction. While opacity can vary noticeably, it must be remembered that postmortem changes " especially if demineralizing " may modify the true picture. Perichondral bone proliferation may produce osteophytes of very variable size, and in some cases union with the next vertebra. Joint malformation and displacement are the end stages, especially in osteoarthritis, where bone on bone movement and eburnation can result in much bone loss and marginal remodeling (Figs. 5.39–5.45).

5.4.1
Osteoarthritis

Also known as degenerative joint disease or osteoarthrosis, this is by far the commonest form of joint disease in mammalian species, and to some extent is age related. The vertebral column and larger joints are most noticeably affected (Fig. 5.39). The development of rims of new bone marginal to joints is typical, with tongue-like osteophytes (osteophytosis) characterizing vertebral changes. Added to these changes are joint-surface pitting and remodeling, together with eburnation (polishing from bone-on-bone movement). X-rays will assist in discriminating between this and other arthropathies. While in zooarcheologi-

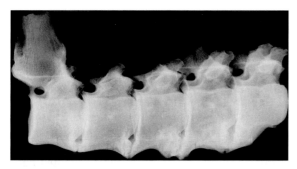

Fig. 5.39. Vertebral osteophyte development in osteoarthritis (slight-medium)

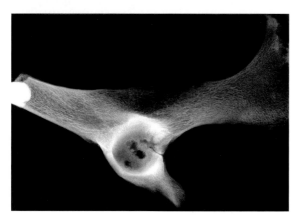

Fig. 5.40. Probable subchondral cysts in a cow pelvis of Iron Age date

cal material, vertebral changes are most common, it does occur noticeably in the stifle joint, especially in dogs (Tirgari and Vaughan 1975). The value of x-rays in differential diagnosis is to reveal any unexpected internal detail and perhaps to discriminate between osteophyte development and the florid projections of bone that can occur in stages of suppurative arthritis (Doige 1980).

Enthesophyte formation, linked to the ossification of soft-tissue insertions, is also distinctive. Subchondral cyst formation may also be associated with osteoarthritic changes (Fig. 5.41). In the field of zooarcheology, it should be noted that there is currently special interest in collecting data on osteoarthritis from the point of view of its relevance in indicating the use of animals for riding or traction purposes (Johannsen 2005; Levine et al. 2005). Such evidence would be relevant in considering the antiquity of such animals as horses, cattle, and camelids for such purposes.

5.4.2
Osteochondritis Dissecans

This is the result of abnormal cartilage development or cartilage trauma, leading to the death of a cartilage "island." This material degrades the joint and may partly calcify, as well as causing remodeling of bone at the joint surface. An x-ray will establish the internal limitations of the changes. Osteochondrosis appears to be a variant, and a more generalized skeletal disturbance (May 1989), resulting from defective endochondral ossification. There is a need for a critical survey of these lesions in mammals.

5.4.3
Legg-Perthes Disease

This may be generally restricted to canids (especially small dogs today) and humans, and results in an aseptic necrosis of the femoral head. This leads to collapse and remodeling of the femur head and neck, with compensatory changes to the pelvic acetabulum.

5.4.4
Infectious Arthritis

Any loss of a smooth joint surface should be questioned as potentially of infectious origin. There will be osteolucent signs of bone destruction, with increasing osseous opacity (Fig. 5.41). Joint surfaces can become highly irregular and can collapse, with subsequent deformity. Areas of florid (nonosteophytic) bone may

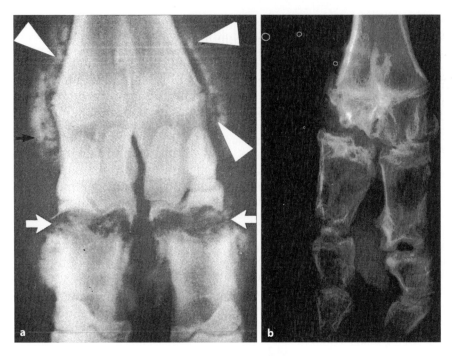

Fig. 5.41 a, b. **a** Young calf with septic arthritis. Joint destruction (arrows) and soft tissue calcification both occur (arrowheads; after Bargai et al. 1989). **b** Healing after septic arthritis in a recent calf (after Douglas and Williamson 1975)

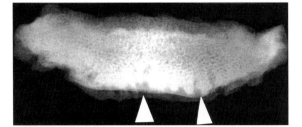

Fig. 5.42. Navicular bone radiograph, showing small multiple invaginations due to the progressive changes in navicular disease

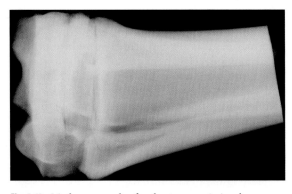

Fig. 5.43. Modern example of early stage spavin in a horse

occur near the joint surface. Radiographic appraisal can reveal the depth and extent of the changes.

5.4.5
Rheumatoid Arthritis

This erosive polyarthritis affects mainly the joints of the extremities, and is reported in dogs and cats, although very uncommonly. It is a progressive condition, with bone changes in the later stages of the disease. Radiographic changes are important in confirming the condition, and include decreased bone opacity at the affected joints, with variable bone destruction and cyst formation. There can also be mushrooming of the articular margins in the metacarpals and metatarsals in the advanced state, together with some degree of joint subluxation and luxation. Additional changes more similar to osteoarthritis may also occur (Allan 2002). While some forms of arthropathy have been described in ancient remains, this particular condition has yet to be identified.

5.4.6
Ankylosing Arthritis

This is really a composite arthritis, probably etiologically complex. It is seen in a variety of mammals, large and small, and has been described in various zooarcheological remains, especially of the horse. In advanced stages, additional bone may seem to "flow"

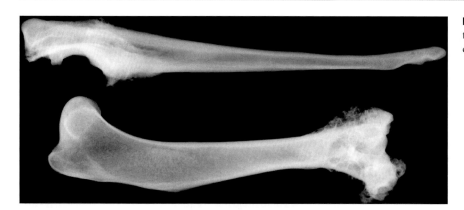

Fig. 5.44. Severe osteoarthritis in the elbow joint of a dog. Much extra bone

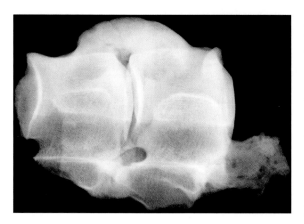

Fig. 5.45. Cow vertebral bodies, displaying large osteophytes. Dragonby, UK

over the surfaces of the vertebral bodies, uniting several of the bones. There can also be the welding together of the neural arches, both involving the articular facets and the bases of the neural spines (Stecher and Goss 1961).

5.4.7
Navicular Disease

The navicular bone is in fact the sesamoid of the third phalanx of the horse (Fig. 5.42). As in the case of other sesamoids at other joints, its job is to minimize friction at the "coffin joint." Trauma in this area may lead to inflammation of the navicular and, if it spreads into the joint, long-term lameness may result (Adams 1979). Evidence of disease in this small bone may therefore indicate that the horse was no longer a healthy and useful riding and traction animal. The bone is not only small, but also complex in shape, and only by radiography can the internal changes to the bone be seen. There is still debate as to the cause of the changes and the nature of the pathology, but

there are clear bone changes nevertheless. It is usually bilateral in the forelimb, and only occasional in the hindlimb. The major radiographic changes are shown in Fig. 5.42, and basically consist of increasing invaginations and remodeling into the distal border, leading to the development of cyst-like lesions.

5.4.8
Bovine Spavin

This is a special form of osteoarthritis that can progress to become an ankylosing arthropathy (Adams 1979). It usually affects the proximal end of the third metatarsal and third and central tarsal bones (Fig. 5.43). Some joint variation can occur. The cause is usually due to poor conformation at the hocks, or of trauma. By radiography, the extent of the actual joint changes and joint ankylosis can be assessed. While usually a condition of horses, it is regarded by some as having an equivalent in cattle, where it involves primarily the central, second, and third tarsal bones. While not uncommon in horses, a similar condition in cows is mostly commonly seen among bullocks (Greenough et al. 1972). The causes in this species may be abnormal conformation and limb stance, as well as excessive stress.

5.5
Neoplasms

(Figs. 5.46–5.55)Relatively little is known of the incidence of tumors in wild populations of vertebrates, but as in human groups, domestic species have been studied in far greater detail. Also, as in our species, it is the malignant forms that are the special concern of veterinarians. For a good general reference text on this disease group, "Tumours in Domestic Animals" is recommended (Moulton 1990). While most cytological features cannot be assessed in archeological material, it is nevertheless possible to tentatively identify benign from malignant tumors (Table 5.2). Both forms may lead to the removal or increase in bone, but there are usually other morphological characteristics that will distinguish the two types (Table 5.2). Benign tumors are structurally well differentiated, slowly growing, and without metastases, although they can become malignant. Malignant forms are usually highly invasive and expansive, with continuous destructive growth and the formation of secondaries. Examples of both benign and malignant tumors are given in Table 5.3, but it must be emphasized that accuracy of diagnosis is greatly reduced when considering incomplete, dry-bone pathology. Moreover, as postmortem burial influences may erode bones and teeth, causing pseudopathology, it is important to distinguish antemortem changes from diagenetic changes. Characteristic changes seen in radiographs may assist in establishing the identity of genuine tumors (Ling et al. 1974), as indeed may the occurrence of new bone in the form of a Codman's triangle (Douglas and Williamson 1975) (Fig. 5.46). Because of the difficulties of a differential diagnosis in archeological material, it is especially important to describe the pathology as accurately as possible, both in morphological and metrical terms, supported by the radiographic evidence. The latter technique may be essential in establishing the occurrence of metastatic deposits, seen usually in bone as irregular zones of destruction at a surface level or deep within bone. Domestic species do not equally develop tumors, and in dogs, some modern breeds are far more susceptible than others. The occurrence of metastases also varies, and in dogs may result in 17% of bone changes (Moulton 1990; Owen 1969).

Probably in all domestic species, if not also in wild forms, the frequency of benign and malignant tumors increases with age. Age at slaughter will of

Table 5.3. Some tumors that can produce bone changes in domestic species

Benign	Malignant
Osteoma	Osteosarcoma
Chondroma	Chondrosarcoma
Osteochondroma	Synovial sarcoma
Giant cell tumor	Liposarcoma
Histiocytoma	Histiocytic sarcoma
Hemangioma	Hemangiosarcoma
Fibroma	Fibrosarcoma
Hemangioma	Myeloma (multiple)
Odontoma	Melanoma, malignant
	Meningioma
	Metastatic (secondary deposits) can result from mammary adenocarcinoma, etc.

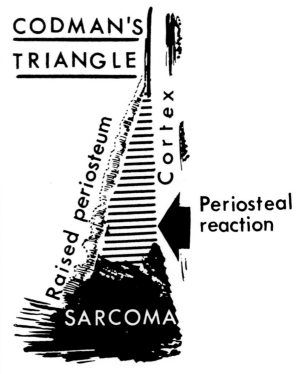

Fig. 5.46. Diagram of Codman's triangle, of neoplastic relevance (after Douglas and Williamson 1975)

Table 5.2. Summary of x-ray contrasts between benign and malignant tumors (modified from Bains 2006)

	Benign	→	Malignant/aggressive
Lysis	Geographic	Moth-eaten	Permeative
Periosteal reaction			
Transition zone	Smooth	Irregular, speculated	Amorphous
Cortical disruption			
Rate of change	Sharp, distinct, short		Indistinct, long
	None		Loss of cortex
	No change or slow		Rapid

course influence the frequencies found. Where animals achieve old age, as in dogs, cats, and horses, the chances of noting tumors may be increased. In sheep, goats, cattle, and pigs, which have a more restricted life expectancy, the chances of noting tumors may be lower. However, some tumors strike at an earlier age, for instance nephroblastoma in young pigs and skin histiocytoma in dogs.

A wide range of species is susceptible to tumors of the skin, with melanomas occurring particularly where pigmentation is heavy. In contrast, primary pulmonary tumors are rare in domestic species. Ingestion of bracken in some countries results in a high incidence of esophageal and stomach cancer in cattle. Mammary tumors are frequent in dogs, less so in cats, and uncommon in other species. Tumors of the nervous system are generally uncommon in domesticates. Some of these tumors produce secondaries in the skeleton. Specific tumors of bone also occur, and in dogs the osteosarcoma is most frequent (Owen 1969). It is important to keep in mind that there may be a hereditary background to tumor development, at least in some species; purebred and inbred dogs display relatively high frequencies. Bovine and equine breeds are affected to a lesser degree. Do inbred varieties of domestic species in prehistory display equally enhanced frequencies of tumors? The diagnosis of neoplasms in domesticates from the past is clearly an important part of biological reconstruction, with radiologic aspects being of considerable importance.

5.5.1
Examples of Tumors Affecting the Skeleton

In contrast to our growing knowledge of tumors in earlier human remains, there is an urgent need for more neoplastic evidence in other species. Some types, such as osteosarcoma in dogs (Fig. 5.47), may well be present in the numerous ancient dog remains, but as yet eludes diagnosis. Others may be geographically restricted, as in the case of carcinoma of the horn core in cattle, which is mainly restricted to India today.

5.5.1.1
Synovial Sarcoma of Joints

While a number of tumors may be sited at joints, this malignant tumor, which especially affects dogs, can be bone destructive. The major weight-bearing joints of the legs are usually affected, and lameness follows. Joint destruction can be extensive, and cystic cavities become well defined at the joints. The discrete borders produced by the osteolysis contrast with bone destruction produced by osteomyelitis, but can be confused with arthritic changes. In contrast to the

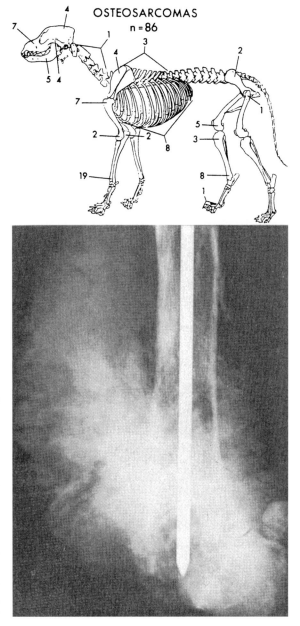

Fig. 5.47. Distribution within the dog skeleton of a sample of osteosarcoma cases. Modified from Ling et al. (1974)

above pathology, specific tumors of bone in mammals are variable from species to species. In dogs, osteosarcomas are far more common than benign conditions, whereas in cattle and horses, benign tumors exceed sarcomas (Pool 1990). In comparison with other organ systems, however, neoplasms of bone are not common, which means that the chances of detecting tumors in archeological material are not great. They do occur, however, and clearly it is important to be aware of their skeletal pathology in order not to miss the evidence in zooarcheological material. Radiographic evidence, in single or multiple views, may provide critical evidence of changes at the borders of lesions or remodeling deep within the architecture of bones. Bone changes may vary from destructive (lytic) to proliferative (sclerotic). Radiographically, the margin of a benign tumor is likely to be smoother and more well defined, while an aggressive malignant neoplasm is likely to produce ragged, poorly defined borders and irregular new bone. The x-rays should confirm the nature of the bone changes as neoplastic, but of course the exact nature of the tumor may remain a very tentative diagnosis. While the zooarcheologist should be proficient in at least separating out pathology from the overall bone sample, and even suggesting the possibility of a neoplasm, discussion with veterinary colleagues is ideal prior to publication. Only the zooarcheologist will have experience of identifying pathology from fragmented bone, and of the variable pseudopathology that can occur as a result of different burial environments.

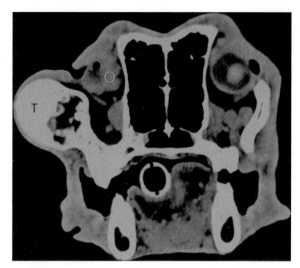

Fig. 5.48. CT image through the nasal area of a cat, to show a large benign osteoma in the zygomatic area. After Johnson and Watson (2000)

5.5.1.2
Benign Tumors

Although in human paleopathology, reports of benign tumors, especially osteomas, are not uncommon, there is far less evidence in other species. Admittedly, in a small sample of neoplasm records for sheep, 2.6% were benign osteoma (Marsh 1965), but in a larger pooled sample the incidence was far smaller.

Osteoma

Smooth, dense and usually solitary bone growth, ranging from small mounds to substantial masses. They appear to favor herbivores, and especially occur in regions of the skull. Radiographically, they are dense bone, merging into the normal cortical bone. They can grow in skull sinus spaces, so that without breakage or x-rays, they can be missed. Occasionally, they can be massive (Fig. 5.48), or multiple, as in the case of an equine cannon bone from Westbridge Friary (Fig. 5.49).

Osteochondroma

These are single or multiple endochondral ossifications, usually found in dogs and to a lesser extent in horses, but other species can be involved. In x-rays, these benign tumors display clear contours, with spongy bone grading into the normal bone (Fig. 5.50). These tumors can become malignant. It should be noted here that enchondromas and chondromas, both cartilage-linked benign tumors, result in bone lesions, but are uncommon. Also, in contrast to humans, hemangiomas are rare in domesticates, but if

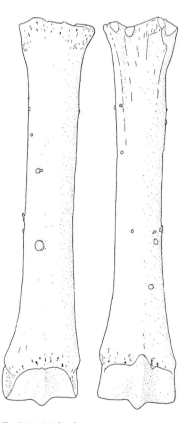

Fig. 5.49. Multiple osteomas on an equine cannon bone from Westbridge Friary, UK. Only CT scans would provide structural detail. Drawing by Clare Thawley)

suspected, then the radiographic picture should be of osteolytic damage, sometimes with an expansive periosteal response in the skull.

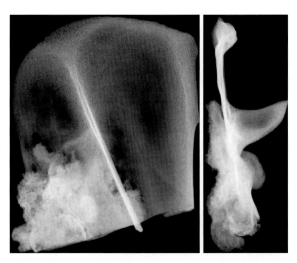

Fig. 5.50 a, b. Osteochrondroma on the scapula of a recent dog

5.5.1.2
Malignant Tumors

Destructive and life-threatening tumors are the most significant from the point of view of interpretation in zooarcheology.

Osteosarcoma

This is a primary bone tumor that is especially common in dogs. For this reason, it would be very interesting to find cases in earlier dog varieties. Radiographic appearances show a poorly delimited lesion with no sclerotic border at the margins. The original cortical surface is replaced by a periosteal response that can produce a considerable mass of spicular bone in a "sunburst" pattern (Fig. 5.51), although this is a variable feature and can be obscured by postburial changes. There is some overlap in the radiographic appearance of osteosarcoma, chondrosarcoma, and fibrosarcoma of bone, but osteosarcoma is most likely to occur in early dogs, as the other forms are relatively rare. This tumor does not cross joints.

Multilobular Tumor of Bone

These progressively malignant, evenly contoured, solitary tumors have yet to be described archeologically, but are the commonest bone tumor of the dog's skull, and thus could well appear in zooarcheological samples. The x-rays may display nodular lesions that can grow to a substantial size and display increased radiodensity and a somewhat granular texture.

Myelomatous Tumors

These are malignant tumors initially of the bone marrow. They are essentially multicentric lytic lesions,
sometimes with a degree of bone proliferation. A possible case has been described in a Roman chicken from Lankhills in Winchester, England (UK), where modest bone proliferation of new bone in the sacrum is associated with multiple lytic lesions in the pelvic basin (O'Connor and O'Connor 2006), as revealed by digital radiography (Fig. 5.52).

Other Tumors Affecting Bone

Although other neoplasms that produce bone pathology are described in the veterinary literature, such as giant cell tumor and liposarcoma, other varieties tend to be rare or sufficiently uncommon as to exclude them from discussion here.

5.5.2
Secondary Tumors of Bone

There is a range of tumors that originates in soft tissues, but may secondarily affect bones. Also, benign forms of bone tumor can be transformed into malignant forms.

5.5.2.1
Metastatic Deposits

Malignant melanomas, mammary carcinoma, and other forms, may by hematogenous metastasis result in the development of both osteolytic and osteoblastic tumors in the skeleton, but this is not common (Fig. 5.53–5.55). Unfortunately, these secondaries do not produce a pathognomonic radiographic appearance. All one can do in the case of zooarcheological material is note any multiple ragged lytic lesions or indistinctive additional bone growths and attempt to discriminate these lesions from those other tumor types described here.

5.5.2.2
Hypertrophic Pulmonary Osteoarthropathy

Hypertrophic pulmonary osteoarthropathy (Marie's disease) in dogs, as in humans, may show distinctive bone deposits that are associated with pulmonary metastases (Fig. 5.54). The bone changes can also occur as a result of other lung disease, so it is not pathognomonic to lung malignancy. Indeed, one of the best archeological cases, in a prehistoric dog from Canada, was shown to have mycobacterial aDNA in the bone lesion, suggesting pulmonary tuberculosis in this case (Bathurst and Barta 2004). The radiographs of such cases display a patchy skeletal spread of subperiosteal new bone, which can thicken out in an "onion peel" layering. Excavations in Colchester, England (UK), have produced two pig bones that are also suggestive of this condition (Luff and Brothwell 1993).

Fig. 5.51. Irregular, but somewhat spicular bone development, on the humerus of a dog, caused by an osteosarcoma

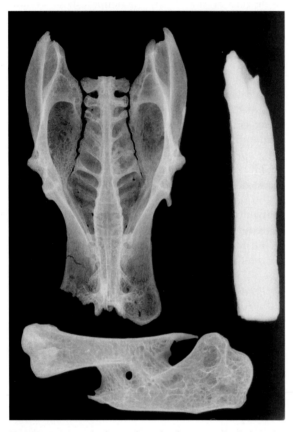

Fig. 5.52. Roman chicken pelvis, displaying multiple lytic lesions, indicating myelomatous tumors. Courtesy of Sonia and Terry O'Connor

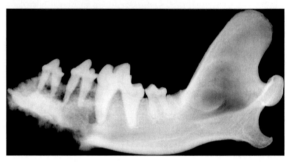

Fig. 5.53. Dog mandible with regional destruction, caused by a metastatic carcinoma

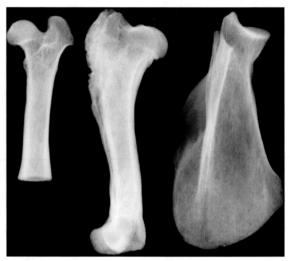

Fig. 5.54. Layered subperiosteal new bone in a dog with Marie's disease

5.6
Oral Pathology

One of the most commonly occurring fields of pathology in zooarcheology is concerned with the jaws and teeth, and is often caused by diet. It may be thought that abnormal conditions of this part of the skull would be perfectly obvious and not in need of more detailed radiographic study. However, use of x-rays may often clarify the extent of the pathology, especially in the interior of jaws, as for instance in the case of deeply impacted teeth, or apical abscesses or early actinomycosis (Figs. 6.56–6.59).

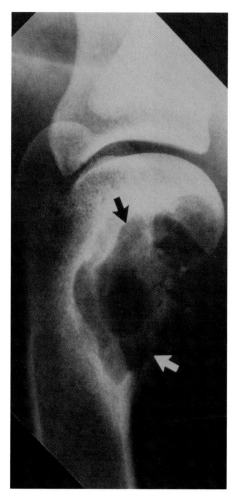

Fig. 5.55. Proximal humerus of a young dog, with bone destruction caused by a chondrosarcoma. After Morgan (1988)

5.6.1
Classifying Oral pathology

5.6.1.1
The Teeth

Where teeth are of normal size and shape in zooarcheological material, then radiography has little value. But there are many anomalies (Miles and Grigson 2003) where x-rays may reveal more detail. These can be summarized as follows:

1. Abnormal position of teeth, buried partly or completely in the jaw. In the case of total burial, this can be missed, but in all cases where a tooth appears to be congenitally absent, an x-ray check is called for.

2. Teeth may be deformed or even fused together, and the extent of the deformity at root level requires radiography.

3. Angled impaction of one tooth against another may need clarification as to the cause.

4. Severe caries, severe attrition or crown trauma, may result in root fragments remaining within the jaw and well below the gum line.

5. Because caries is uncommonly found in nonhuman teeth, estimating the extent of the decalcification by x-ray would be worth while. However, it is important to be aware of pseudocaries, caused by postmortem diagenetic factors.

6. Severe wear leading to pulp exposure usually leads to apical infection, which may only be detected in radiographs.

5.6.1.2
The Jaw Bones

In young animals, the jaws are usually in a healthy state, unless trauma and related infection have left their mark. The other exception is craniomandibular osteopathy of dogs, as yet not noted in archeological material (Johnson and Watson 2000) (Fig. 5.56). This is a distinctive proliferative bone disease occurring in young animals and producing a swelling especially at the base of the mandibular body. There is increased bone opacity in the swollen area. It may be bilateral but not symmetrical. Other more common conditions may be listed as follows.

1. Periodontal disease consists of the infection of various tissues that support or are associated with the teeth (DeBowes 2000) (Fig. 5.57). This in particular involves the alveolar bone surrounding the tooth roots, which displays resorption and recession, exposing the roots and causing increasing looseness of teeth. This is to some extent wear- and age-related. It can also be influenced by calculus (tartar) development, which can stimulate local or massive periodontal changes. Microtrauma from coarse cereal foods may also cause periodontal infection and eventual tooth loss. The extent of the alveolar bone destruction, especially between the teeth and into deeper parts of the tooth socket, can best be revealed by x-ray.

2. Apical infection. As a sequel to severe caries, trauma, or as a result of attrition-related pulp exposure, infection may be established in the bone surrounding the root apex. Bone destruction results in a rounded bony cavity, an abscess, and finally leads to loss of one or more teeth.

3. Apical infection (root abscess). These may not be obvious without an x-ray of the jaw. When teeth are lost, the abscess chamber may be examined visually, but can often extend under teeth still in place.

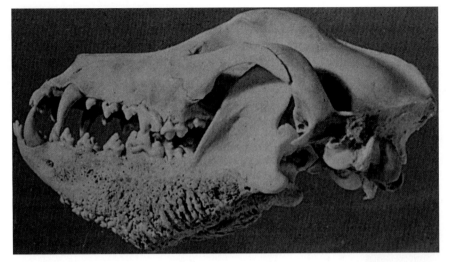

Fig. 5.56. The enlarged lower jaw of a dog with cranio-mandibular osteopathy. After Johnson and Watson (2000)

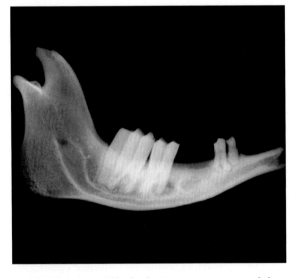

Fig. 5.57. Sheep mandible displaying ante-mortem tooth loss, tooth drifting, and some alveolar recession

4. Antemortem tooth loss (Fig. 5.58). This follows usually from the processes discussed above. However, it can be the result of accidental trauma, and it deserves to be noted that poor handling by humans in earlier cultures might well have caused tooth loss.

5.6.1.3
Comparative and Epidemiological Studies

As oral pathology is one of the most commonly occurring fields of pathology in both recent and ancient material, diagnosis and recording of the conditions might potentially enable comparison between sam-

ples of different periods and regions. The range of abnormalities and the different jaw and tooth positions that can be affected are illustrated by two Pleistocene American lions (*Panthera leo atrox*) from the Yukon territory (Beebe and Hulland 1988). Between them, the specimens display evidence of chronic periodontitis, osteomyelitis, congenital absence of incisors, and antemortem loss of a lower canine. Not only did x-rays reveal tooth root detail, but indicated clearly that in one mandible, the original cortex of the mandibular body was intact, even though the area was swollen. Also, in the other mandible, the socket for the right canine is completely infilled with bone.

Periodontal infections and tooth loss are common in the majority of mammals when they become old and tooth wear is severe, as evidenced by great apes and sheep (Dean et al. 1992; Newton and Jackson 1984). In view of the commonness of such pathology, perhaps especially in cave bears in the Pleistocene, it is surprising that detailed surface and radiographic studies have not been undertaken on bears (Bachofen-Echt 1931; Moodie 1923) and other ancient species, but this remains an interesting research challenge for the future.

Precise evaluation of oral pathology demands radiographic checks for another reason. There is clearly regional variation in the frequencies of these pathologies, as exemplified by studies on wild pigs from Israel (Horwitz and Davidovitz 1992). This may be partly due to inbreeding, but some pathologies are also influenced by environmental contrasts.

The regular occurrence of oral pathology cannot be too strongly emphasized. A visual and radiographic examination of the jaws of 581 recent adult culled ewes, revealed that only two were considered to be normal (Richardson et al. 1979). X-rays revealed bone rarefaction, poor root development, and pro-

Fig. 5.58. Occlusal view of a sheep mandible, displaying tooth loss and apical abscessing. Medieval

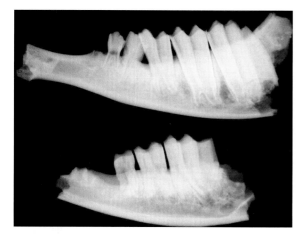

Fig. 5.59. Iron Age sheep mandible, displaying tooth impaction and the drifting of teeth. Dragonby

found cemental hyperplasia. Yet other mandibular pathology was seen on a Scythian horse, which displayed additional bone "excrescences," interpreted as the result of chronic mechanical irritation from harnessing (Bokonyi 1968).

The need to develop a methodology to record oral pathology as precisely as possible has been discussed to some extent for instance in relation to "broken mouth" and periodontal disease in sheep (Spence et al. 1980). Similarly, recording methods are discussed in relation to ancient remains by Levitan (1985) and Baker and Brothwell (1980). Clearly, radiographic studies deserve an input into any such methodology.

5.7 Conclusions

I have tried to draw together in this chapter a range of diseases where radiography has a part to play in aspects of study and diagnosis. Textbooks on veterinary pathology and radiology are often not as comprehensive as they could be, and it is hoped that this chapter has no intentional pathology or species bias, except that it has excluded humans from considering mammals. The literature on veterinary skeletal and dental pathology is vast, and it has been necessary to strictly control the bibliographic list. Our hope is that

this work is comprehensive enough to be used as an introductory reader in the field of paleoradiology, including veterinary paleopathology.

The radiography of archeological material has both advantages and disadvantages. We don't normally have the whole skeleton, and rarely a well-preserved mummified body with other tissue. On the other hand, the dry-bone pathology, combined with x-rays, can be very revealing, even when bones are incomplete. Most importantly, this is a largely unexplored field with great potential.

References

Adams OR (1979) Lameness in Horses, 3rd edn. Lea and Feibiger, Philadelphia

Allan G (2002) Radiographic signs of joint disease, In: Thrall DE (ed) Textbook of Veterinary Diagnostic Radiology 4th edn. Saunders, Philadelphia, pp 187–207

Andrews AH (1985) Osteodystrophia fibrosa in goats. Vet Annual 25:226–230

Bachofen-Echt A (1931) Abnorme zahnstellung bei keifern von Ursus deningeri aus Mosbach. Palaeobiologica 4:345–351

Baines E (2006) Clinically significant developmental radiological changes in the skeletally immature dog: 1. long bones. Practice 28:188–199

Baker JR, Hughes IB (1968) A case of deforming cervical spondylosis in a cat associated with a diet rich in liver. Vet Rec 81:44–45

Baker J, Brothwell D (1980) Animal Diseases in Archaeology. Academic Press, London

Bargai U, Pharr JW, Morgan JP (1989) Bovine Radiology. Iowa State University Press, Ames

Bathurst RR, Barta JL (2004) Molecular evidence of tuberculosis induced hypertrophic osteopathy in a 16th-century Iroquoian dog. J Archaeol Sci 31:1–9

Beebe BF, Hulland TJ (1988) Mandibular and dental abnormalities of two Pleistocene American lions (*Panthera leo atrox*) from Yukon territory. Can J Vet Res 52:468–472

Bokonyi S (1968) Mecklenburg Collection, Part 1: Data on Iron Age Horses of Central and Eastern Europe Bull 26, Peabody Museum, Harvard University, Cambridge

Boulay GH du, Crawford MA (1968) Nutritional bone disease in captive primates. Symp Zool Soc Lond 21:223–236

Brothwell D (1979) Roman evidence of a crested form of domestic fowl, as indicated by a skull showing associated cerebral hernia. J Archaeol Sci 6:291–293

Brothwell D (1995) The special animal pathology. In: Cunliffe B (ed) Dancbury, An Iron Age Hillfort in Hampshire, 6: A

Hillfort Community in Perspective. CBA Research Report 102, CBA, York, pp 207–233

Brothwell D (2002) Ancient avian osteopetrosis: the current state of knowledge. Acta Zool Cracov 45:315–318

Brothwell D, Schreve D, Boismier W (2006) On the palaeopathology of mammoths, with special reference to cases from Lynford, England. Quat Int (in press)

Chai CK (1970) Effect of inbreeding in rabbits: skeletal variations and malformations. J Hered 61:3–8

Dean MC, Jones ME, Pilley JR (1992) The natural history of tooth wear, continuous eruption and periodontal disease in wild shot great apes. J Hum Evol 22:23–39

DeBowes LJ (2000) Dentistry: periodontal aspects. In: Ettinger SJ, Feldman EC (eds) Textbook of Veterinary Internal Medicine: Diseases of the Dog and Cat, 5th edn. Saunders, Philadelphia, pp 1127–1142

Dennis R (1989) Radiology of metabolic bone disease. Vet Annual 29:195–206

Doige CE (1980) Pathological changes in the lumbar spine of boars. Can J Comp Med 44:382–389

Douglas SW, Williamson HD (1975) Veterinary Radiological Interpretation. Heinemann, London

Fink MT (1985) Coccidioidal bone proliferation in the pelvis (os coxa) of canids. In: Merbs CT, Miller RJ (eds) Health and Disease in the Prehistoric Southwest. Arizona State University, Tempe, pp 324–339

Fowler ME (1989) Medicine and Surgery of South American Camelids. Iowa State University Press, Ames

Greenough PR, MacCallum FJ, Weaver AD (1972) Lameness in Cattle. Oliver and Boyd, Edinburgh

Harris HA (1931) Lines of arrested growth in the long bones in childhood. Br J Radiol 4:561–588, 622–640

Holmes JR, Baker JR, Davies ET (1964) Osteogenesis imperfecta in lambs. Vet Rec 76:980–984

Horwitz LK, Davidovitz G (1992) Dental pathology of wild pigs (Sus scrofa) from Israel. Israel J Zool 38:111–123

Jeffcott LB, Whitwell KE (1976) Fractures of the thoracolumbar spine of the horse. Proceedings of the 22nd Annual Convention of the American Association of Equine Practitioners, Dallas, pp 1–12

Jensen R, Mackey DR (1979) Diseases of Feedlot Cattle. 3rd edn. Lea and Feibiger, Philadelphia

Johannsen NN (2005) Palaeopathology and Neolithic cattle traction: methodological issues and archaeological perspectives. In: Davies J, Fabis M, Maintand I, Richards M, Thomas R (eds) Diet and Health in Past Animal Populations: Current Research and Future Directions. Oxbow, Oxford, pp 39–51

Johnson KA, Watson AD (2000) Skeletal diseases. In: Ettinger SJ, Feldman EC (eds) Textbook of Veterinary Internal Medicine: Diseases of the Dog and Cat, 2, 5th edn. Saunders, Philadelphia, pp 1887–1916

Jubb KV, Kennedy PC, Palmer N (1985) Pathology of Domestic Animals. 3rd edn. Academic Press, London

Kennedy B, Moxley JE (1980) Genetic factors influencing atrophic rhinitis in the pig. Anim Prod 30:277–283

Leipold HW (1997) Congenital defects of the musculoskeletal system. In: Greenough PR, Weaver AD (eds) Lameness in Cattle. Saunders, Philadelphia, pp 79–85

Levine MA, Whitwell KE, Jeffcott LB (2005) Abnormal thoracic vertebrae and the evolution of horse husbandry. Archaeofauna 14:93–109

Levitan B (1985) A methodology for recording the pathology and other anomalies of ungulate mandibles from archaeological sites. In: Fieller NRJ, Gilbertson DD, Ralph NG (eds) Palaeobiological Investigations. BAR (Internat) 266, Oxford, pp 41–54

Ling GV, Morgan JP, Pool RR (1974) Primary bone tumors in the dog: a combined clinical, radiographic, and histologic approach to early diagnosis. J Am Vet Med Assoc 165:55–67

Luff R, Brothwell D (1993) Health and welfare. In: Luff R (ed) Animal Bones from Excavations in Colchester, 1971–85. Colchester Archaeological Report, 12: Colchester Archaeological Trust, Colchester, pp 101–126

Maddy KT (1958) Disseminated coccioidomycosis of the dog. J Am Vet Med Assoc 132:483–489

Marsh H (1965) Newsom's Sheep Diseases. 3rd edn. Williams and Wilkins, Baltimore

May C (1989) Osteochondrosis in the dog: a review. Vet Ann 29:207–216

Mays SA (2005) Tuberculosis as a zoonotic disease in antiquity. In: Davies J, Fabis M, Mainland I, Richards M, Thomas R (eds) Diet and Health in Past Animal Populations: Current Research and Future Directions. Oxbow, Oxford, pp 125–134

Miles AEW, Grigson C (2003) Colyer's Variations and Diseases of the Teeth of Animals. Cambridge University Press, Cambridge

Moodie RL (1923) Paleopathology: an Introduction to the Study of Ancient Evidences of Disease. University of Illinois Press, Urbana

Morgan JP (1988) Radiology of skeletal disease – principles of diagnosis in the dog. Iowa State University Press, Ames

Moulton JE (ed) (1990) Tumors in domestic animals, 3rd edn. University of California Press, Berkeley

Murphy EM (2005) Animal palaeopathology in prehistoric and historic Ireland: a review of the evidence. In: Davies J, Fabis M, Mainland I, Richards M, Thomas R (eds) Diet and Health in Past Animal Populations: Current Research and Future Directions. Oxbow, Oxford, pp 8–23

Newton JE, Jackson C (1984) The effect of age on tooth loss and the performance of masham ewes. Anim Prod 39:421–425

O'Connor T, O'Connor S (2006) Digitising and image processing radiographs to enhance interpretation in avian palaeopathology. In: Grupe G, Peters J (eds) 2005 Feathers, grit and symbolism, birds and humans in the ancient Old and New Worlds. Leidorf, Rahden, pp 69–82

Owen LN (1969) Bone Tumours in Man and Animals. Butterworth, London

Pales L (1930) Paléopathologie et pathologie comparative. Masson, Paris

Payne LN (1990) Leukosis/sarcoma group. In: Jordan FTW (ed) Poultry diseases, 3rd edn. Bailliere Tindall, London, pp 106–115

Pedersen NC, Morgan JP, Vasseur PB (2000) Joint diseases of dogs and cats. In: Ettinger SJ, Feldman EC (eds) Textbook of Veterinary Internal Medicine, 2, Diseases of the Dog and Cat. 5th edn. Saunders, Philadelphia, pp 1862–1886

Penny RHC, Mullen PA (1975) Atrophic rhinitis of pigs: abattoir studies. Vet Rec 96:518–521

Platt BS, Stewart RJC (1962) Transverse trabeculae and osteoporosis in bones in experimental protein-calorie deficiency. Br J Nutr 16:483–496

Pool RR (1990) Tumors of bone and cartilage. In: Moulton JE (ed) Tumors in Domestic Animals, 3rd edn. University of California Press, Berkeley, pp 157–230

Richardson C, Richards M, Terlecki S, Miller WM (1979) Jaws of adult culled ewes. J Agric Sci Camb 93:521–529

Rogers J, Waldron T, Dieppe P, Watt I (1987) Arthropathies in palaeopathology: the basis of classification according to most probable cause. J Arch Sci 14:179–193

Schultz AH (1939) Notes on diseases and healed fractures of wild apes. Bull Hist Med 7:571–582

Spence JA, Aitchison GU, Sykes AR, Atkinson PJ (1980) Broken mouth (premature incisor loss) in sheep: the pathogenesis of periodontal disease. J Comp Path 90:275–292

Stecher RM, Goss LJ (1961) Ankylosing lesions of the spine. J Am Vet Med Assoc 138:248–255

Tasnadi Kubacska A (1960) Az Osállatok Pathologiája. Medicina Konyukiado, Budapest

Tirgari M, Vaughan LC (1975) Arthritis of the canine stifle joint. Vet Rec 18:394–399

Toal RL (2002) The navicular bone. In: Thrall DE (ed) Textbook of Veterinary Diagnostic Radiology, 4th edn. Saunders, Philadelphia, pp 295–305

Toal RL, Mitchell SK (2002) Fracture healing and complications. In: Thrall DE (ed) Textbook of Veterinary Diagnostic Radiology, 4th edn. Saunders, Philadelphia, pp 161–178

Udrescu M, Van Neer W (2005) Looking for human therapeutic intervention in the healing of fractures of domestic animals. In: Davies J, Fabis M, Mainland I, Richards M, Thomas R (eds) Diet and Health in Past Animal Populations; Current Research and Future Directions, Oxbow, Oxford, pp 24–33

Vlcek E, Benes J (1974) Über eine schädelassymetrie der höhlenhyäne Crocuta spelaea (Goldfuss), die als folge einer unilateralen schädelverletzung entstanden ist. Lynx 15:31–44

CHAPTER 6

Normal Variations in Fossils and Recent Human Groups

6

DON R. BROTHWELL

What is normality in terms of variation seen in fossil remains or the large skeletal samples excavated from sites of the last 10,000 years? There tends to be an assumption that we know what the boundaries of normality are, and thus "abnormality" presents no problems in terms of its differentiation. But this seems to be a matter for some debate, and intrapopulation studies on biological variation in skeletal and dental remains are by no means common. In particular, variation revealed by radiographic study is so far poorly reported in the literature. Moreover, it is probably true to say that variation in fossil humans has been especially neglected, and this applies to revealing and confirming pathology as well as establishing normal variation. A few examples will establish the ways in which x-rays could have assisted in extending our paleontological knowledge. Take for instance the East African skulls KNM-ER 406, KNM-ER 1470, and KNM-ER 1813 (Leakey et al. 1978), x-rays and computed tomography (CT) scans would have provided important extra information about cranial thickness, size and shape of frontal, maxillary, and mastoid sinuses, and perhaps even information on the basicranial angle and size and shape of the sella turcica. In the same way, x-rays were needed to fully appreciate the morphology and degree of breakage and distortion of the Arago XXI skull (de Lumley 1981).

In the case of complete and unbroken skulls, other aspects of the inner architecture could potentially be revealed by radiography. For instance, Tobias (1968) has shown differences in the venous sinus grooves in the posterior cranial fossa of *Australopithecus boisei* and Swartkrans 859, which could be identified radiographically. Yet again, another form of enquiry to benefit from the application of radiological techniques would be cranial deformation. In the case of the Kow Swamp series (Thorne 1971), I believe that lateral x-rays would have greatly assisted in revealing clearly the extent of bone modification, as well as providing information on cranial thickness and facial conformation.

Could the inner architecture as revealed by x-rays also assist in deciding on the correct head positioning of fossil skulls? In the case of the Chinese Mapa cranial fragment (Woo and Peng 1959), additional information on the endocranial surface of the frontal and its sinus system could have helped to establish the correct orientation into the Frankfort horizontal. As regards the postcranial skeleton, accuracy of measuring the femoral neck-shaft angle would be better achieved from x-rays, especially as so often the fossil remains can be damaged (as in the Qafzeh-Skhul material; Trinkaus 1993).

Finally, of course, all fossil pathology should be x-rayed. For instance the pitting on the Krapina cranial fragment 34.12 (Radovčić et al. 1988) deserves further exploration, if possible by CT scan. Could the changes be postmortem? This is another aspect of human paleontological studies that could be assisted by radiological evaluation, and in the case of the Swanscombe cranial bones, the nature of the postmortem bone damage (Fig. 6.1) was clearly revealed in a series of x-rays (Le Gros Clark 1964).

6.1
Fossil Studies by Conventional Radiography

Generally, except for the teeth and jaws, radiological studies on fossil hominins have progressed far

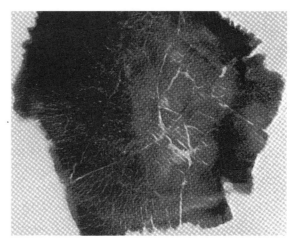

Fig. 6.1. Damage to the right parietal bone of the Swanscombe skull, as revealed by x-ray (Le Gros Clark 1964)

too slowly, and only in recent years has the situation started to change. Gorjanović-Kramberger (1906) was a pioneer in x-raying Neanderthal material from Krapina. The Canadian anatomist Davidson Black (1929) was similarly aware of the value of x-rays, although limiting his studies in this case to the jaws (Fig. 6.2). Further x-ray studies on this *Homo erectus* material were published later by Weidenreich (1935) and others, but again only a restricted amount was radiographed. Surprisingly little then followed on *H. erectus* and was published, although the Trinil femora received detailed attention and somewhat surprisingly displayed no features that could be used to distinguish them from modern femora (Day and Molleson 1973).

The Neanderthal centenary celebration of 1956 resulted in the publication of a series of specialist papers (von Koenigswald 1958), including a comparative study of the Rhodesian (Fig. 6.3), Florisbad and Saldanha skulls (Singer 1958). This included lateral x-rays of the three specimens, and it appeared that the bones of the Florisbad calotte were much thicker than the other two. This raises the question again of the significance of cranial thickness in fossil hominins, including some erectines and Solo specimens. This is very likely to indicate environmental stress (resulting in anemia) rather than being a useful paleontological trait. Regrettably, the earlier x-rays of the Solo skulls (Jacob 1967) did not show internal detail very clearly, owing to the degree of mineralization (Fig. 6.4), and

there is certainly a need to CT scan all the fossil hominins, a task already being gradually carried out.

During this period, Professor J.S. Weiner undertook to x-ray as many human fossils as he could gain access to. Unfortunately, these were never published as an atlas of hominin radiographs, and no critical analysis appears to have been made. The project did, however, help to stimulate interest in the potential data that could be derived from such x-rays. A more detailed study was undertaken on the highly fossilized Rhodesian (Kabwe) skull (Price and Molleson 1974), supporting the diagnosis of an antemortem squamous temporal lesion, but doubting mastoid infection in life. X-rays also supported pathological diagnosis of a fossil parietal from Cova Negra (trauma with infection) and another parietal from Lazaret, with bone changes possibly indicative of a meningioma (Lumley-Woodyear 1973). Trinkaus (1983) similarly employs x-rays to explore the nature of the pathology seen in the Shanidar Neanderthalers.

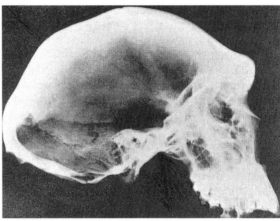

Fig. 6.3. Lateral view of the Rhodesian skull

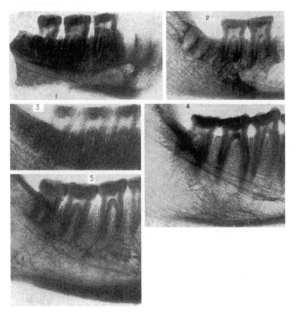

Fig. 6.2. X-rays of a Chinese Homo erectus mandible (1), compared with the Piltdown mandible (2), Heidelberg jaw (3), a recent Chinese (4), and a female adult orangutan (5). From Black (1929)

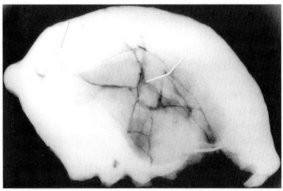

Fig. 6.4. Lateral view of Solo 10, as taken by J.S. Weiner, showing heavy fossilization. Courtesy of the Trustees, Natural History Museum, London

In the Taung Diamond Jubilee symposium proceedings, one section was entitled "new shadow picture beyond Röntgen's wildest visions." The pioneer studies helped to introduce paleontology to the merits of CT, and considered such questions as cranial capacity estimates from matrix-filled skulls (Conroy and Vannier 1985), the assessment of intracranial morphology (Zonneveld and Wind 1985), and temporal bone structure and variation (Wind and Zonneveld 1985).

6.2
Teeth and Jaws

The value of x-rays in the evaluation of oral, and especially dental variation, was appreciated early in paleontological studies. Not only were the jaws of *H. erectus* being radiographed (Black 1929; Weidenreich 1935), but distinctive features of the Neanderthalers were clearly visible by x-rays. Thus, excellent detail of the jaws of the Gibraltar child were produced (Fig. 6.5), showing clearly the well-developed taurodonty in the erupted molars (Buxton 1928). Since then, Kallay (1963) and others such as Skinner and Sperber (1982) have successfully used radiography to reveal differences in the internal structure of teeth, from enamel thickness to pulp chamber size (Fig. 6.6). Unfortunately, the nature of fossilization in some material, such as the hominins of Hadar, Sterkfontein, Kromdraai, Taung, Swartkrans, and other sites has produced in some cases a chalky and poorly differentiated tissue appearance. The degree of development of deciduous and permanent teeth can usually be assessed (as in Swartkrans SK63 for instance).

In contrast to the commonly occurring taurodonty of the Neanderthalers, the pulp chambers of *H. erectus*, as exemplified by the Ternifine jaws (Arambourg 1963) and Nariokotome youth (Brown and Walker 1993), appear to be more comparable to modern proportions (Fig. 6.7). This is also the case with advanced Upper Paleolithic communities, as for instance in the Le Placard, Solutrean jaws (Skinner and Sperber 1982).

6.3
The Advent of CT

Perhaps because it represents new technology, CT scanning has attracted far more attention in paleontology than the older conventional x-ray techniques. In fact there is an argument for a combination of old and new. For instance, Harris lines have never been reviewed in a broad range of fossil hominins, but their detection could be assisted by a combination of old and new techniques. In my own experience of exam-

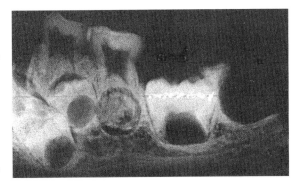

Fig. 6.5. The teeth of the Gibraltar Neanderthal child, as revealed by x-ray (as shown by Buxton 1928)

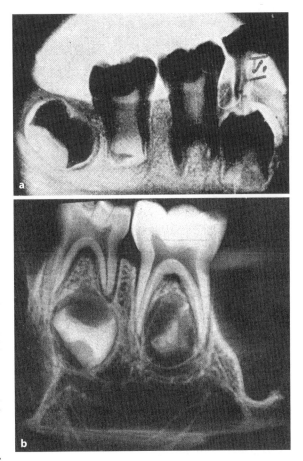

Fig. 6.6. Examples of Krapina Neanderthal (**a**) and Solutream teeth (**b**) in x-rays

ining curated x-rays, La Chapelle aux Saints, Paviland, Wadjak, Shanidar 3, and Tabun 1, appear to show no lines. But there could be old partial lines in Krapina material, as well as Spy 2 and Rhodesian postcranials, so there is a good case for the specific study of Harris lines in fossil material. Other anomalies in fossil material of course also need investigation, and

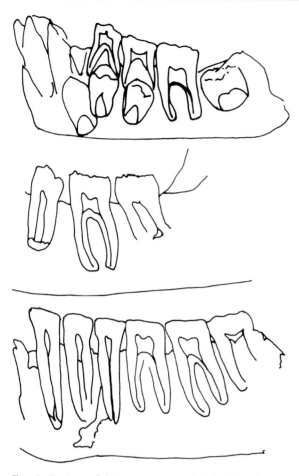

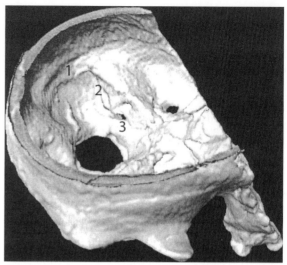

Fig. 6.8. Computed tomography (CT) reconstruction of the australopithecine MLD 37/38, cleaned of matrix, and showing a transverse sinus groove (1), sigmoid sinus groove (2), and jugular foramen (3). From Neubauer et al. (2004)

Fig. 6.7. Tracings of the Nariokotome subadult (*middle*) compared with mandibles KNM-34 (*top*) and KNM-ER 992 (*bottom*). From Brown and Walker (1993)

the possibility of Scheuermann's disease in the Afar australopithecine vertebrae (Cook et al. 1983) would benefit from CT investigation. The application of CT scanning to the study of the anomalous Singa skull was certainly able to reveal that the right temporal bone lacks the structures of the bony labyrinth (Spoor et al. 1998).

Overall scanning of skulls and postcranial bones will in the future clearly be part of the methodology of investigating fossil material. As Seidler and colleagues also point out, the stereolithography of fossil skulls also assists in the comparison of external and endocranial morphology, where there is more variability in structural relationships than has been appreciated in the past (Seidler et al. 1997). Also, three-dimensional techniques enable the selective reconstruction of only parts of a fossil, as in the Broken Hill skull, where both external and some internal detail are associated together (Spoor and Zonneveld 1999). In the same way, three-dimensional information was provided on the partial australopithecine skull MLD 37/38

(Fig. 6.8) by Neubauer et al. (2004). The new CT study of the Le Moustier Neanderthal teenager would also suggest that patterns of differential growth in the face by adolescence is distinctive in this fossil group (Thompson and Illerhaus 2000).

6.4
The Cranial Sinuses

Skull pneumatization has variable functions in mammals, and from the point of view of human evolution, the frontal, sphenoid, maxillary, and mastoid sinus systems have not had equal importance. The mastoid process, if not the temporal sinus system in general, has increased in size during hominin evolution, and is associated with the balance of the head. As Koppe and Nagai (1999) point out, the maxillary sinuses are influenced in their development by factors of diet, masticatory stress, craniofacial growth, dental variables, and malocclusion. Certainly there are big size differences, and the maxillary sinus volume in the Broken Hill skull is twice that of modern humans (Spoor and Zonneveld 1999). As yet there is only limited comparative fossil information. While less can be said of the sphenoid sinus, it is interesting that in the Swanscombe occipital, the basioccipital displays the posterior extension of the sphenoid sinus, which raises the question of sinus correlations, and thus does a large sphenoid sinus suggest that the frontal sinus system was equally large (influencing the form of the supraorbital area)? Clearly experimental studies could help to resolve some of the questions related to variation.

6.5
The Frontal Sinuses

By far the most is known about sinus variation in the frontal bone. Expansion into the frontal from the nasal area occurs particularly in the teenage years and is a sexually dimorphic characteristic. There are probably family differences (Szilvássy 1982), and in any careful evaluation of this kind, it is important that the orientation of the skull (or head) is correct and the same (Schüller 1943). This is also important if the sinus size and shape has forensic relevance (Krogman 1962).

In terms of evolutionary variation, the frontal sinuses in Pleistocene fossils vary considerably. This is seen even within one group, such as the Neanderthalers (Fig. 6.9), only part of the variation being explainable in terms of sexual dimorphism (Vlček 1967). In a brief evaluation of 19 Middle- and Upper-Pleistocene fossils, I found that only 3 (15.8%) displayed very small or small frontal sinuses, while an equal number (42.1%) had medium to large sinuses. Six (31.6%) had marked asymmetry in sinus size. In the case of Holocene populations, the surface areas of the frontal sinuses, as viewed in anteroposterior x-rays, was highly variable and did not seem to be associated with supraorbital size and prominence (Brothwell et al. 1968). Buckland-Wright (1970) considered sinus variation in early British populations from the Bronze Age through to medieval times, and again was able to demonstrate differences between the groups, perhaps influenced both by genetic and environmental factors.

6.6
Variation in Recent Populations

It can be seen from the previous discussion on sinus variation, that some radiological investigations extend from earlier hominins and the possible evolutionary relevance of structural differences, to the microevolutionary or environmentally determined variation of Holocene peoples. In the latter groups, we are dealing with modern physical appearance, but nevertheless with intra- and intergroup distinctiveness. Some fields are relatively well researched, such as dental development, eruption, and pathology, as also is orthodontic variation, although in both cases, there is a need for far more regional and ethnic investigations, and the impact of environmental stress on the rate of progress of skeletal and dental features. In the case of tribal collections, these are being increas-

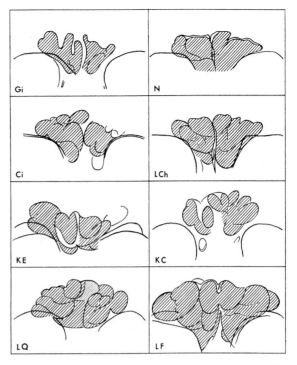

Fig. 6.9. Neanderthal frontal sinus variation, as illustrated by Vlček (1967), from x-rays. *Gi* = Gibraltar, *N* = Neandertal, *Ci* = Cicero, *LCh* = La Chapelle, *KE* = Krapina E, *KC* = Krapina C, *LQ* = Quina, *LF* = La Ferrassie. Original Vlček

ingly returned to local communities, who are being given the power to curate or destroy material that will never be available again. Such is the wisdom of our political masters, that they show little regard for the long-term scientific value of such skeletal material. The only good result of such "repatriation" is that there has been a need to consider the radiographic recording of large series of skeletal remains (Fairgrieve and Bashford 1988) and develop protocols that will take into consideration all aspects of variation (Bruwelheide et al. 2001).

6.7
Age and Growth

Even before the discovery of x-rays, it was known that tooth development and eruption could be used as a guide to age determination in children (Saunders 1837). By means of radiography, more precise information could be obtained on teeth, and over the decades a considerable literature has grown in relation to this topic (Bang 1989). Its relevance clearly extends well beyond the dental profession and dental ageing has been applied to numerous forensic cases, as well as age assessment of children from earlier populations. In comparison with dental attrition and root translucency measurements, dental development remains one of the most reliable of ageing methods, though of course restricted in age range (Whittaker 2000). As regards ethnic and regional variation in tooth development, although there is sufficient evi-

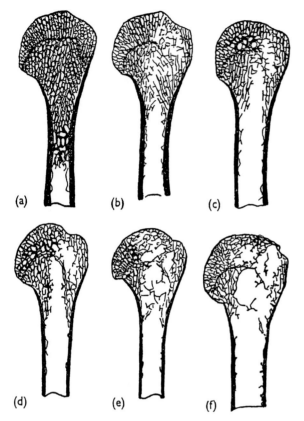

Fig. 6.10. Cancellous tissue changes in relation to age, as proposed by Acsádi and Nemeskéri (1970), employing x-rays

dence to conclude that both within-group and regional (genetic/environmental) differences occur (El-Nofely and Iscan 1989), far more detail is needed, especially in relation to earlier populations. And while Upper Paleolithic communities may display eruption timing relatively similar to modern populations, Neanderthals, *H. erectus* and other fossil hominins may show modified times of development and eruption. Radiological studies of skeletal growth have also helped to provide standards by which to evaluate ethnic variation as well as considering past populations. While data from dry bone specimens have been particularly valuable (Ubelaker 1989), x-rays can be useful in defining the age of some adults (Sorg et al. 1989). In particular, the degree of epiphyseal union and the extent of older adult demineralization (in postmenopausal osteoporosis). Changes in the inner architecture at the articular ends of long bones in relation to age (Fig. 6.10) were schematized by Acsadi and Nemeskéri (1970) and further radiographic studies have since been carried out (Walker and Lovejoy 1985). It can be concluded that radiographic evaluation is of value, but that single age indicators are not as accurate as a combination of multiple factors.

It seems likely that further investigations of craniofacial growth, and the suitability of different orientations of the skull will assist in the correct comparisons of this part of the skeleton, both of living peoples and those in the past. While external cranial morphology is still the main area studied in paleontology and archeology, inner cranial morphology is increasingly seen to be variable and worthy of study (Schuster and Finnegan 1977). Moss (1971) using information on normal and Down's syndrome Danes (Kisling 1966), shows that the inner cephalometric points employed in the orientation of radiographs

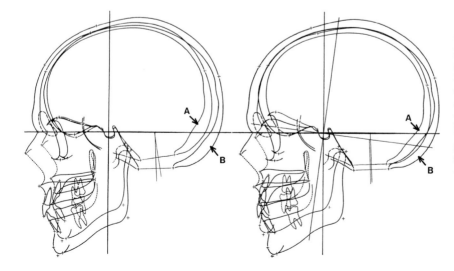

Fig. 6.11. Radiographic contours and cephalometric positioning of two Danish groups: **a** Down's syndrome males; **b** normal males. *Left* contours in the nasion-sella orientation. *Right* contours are positioned on the endocranial base. Relationships between groups change in emphasis as a result. Modified from Moss (1971)

can noticeably change other cranial relationships (Fig. 6.11).

Further ethnic studies of the kind carried out in Australia and Peru will certainly need radiographic support. Brown's study (Brown 1973) revealed significant differences in the inner cranial morphology of tribal Australian samples. While the Peruvian study did not include x-rays, it raises again the importance of this technique in the full evaluation of human biological variation (Pawson et al. 2001). The two communities involved were high-altitude groups, one being in a mining area and showing more rapid skeletal development. However, differences possibly resulting from socioeconomic factors may have also been influenced by high altitude hypoxia, and these complex influences on growth demand the comparison of high and low altitude groups, both now and in the past.

6.8
Sella Turcica Variation

The pituitary fossa is surrounded by part of the sphenoid, and is complex in shape. It is best seen in lateral x-ray, when it appears as a rounded cavity in the endocranial base, with an upper projection of the anterior clinoid process and at the back the posterior clinoid process. The cavity and surrounding bone is variable in normal shape, and may be further modified by disease, especially intracranial tumors. Some variation can be the result of ligament ossification in relation to the clinoid processes. Sella bridging is viewed by some as a nonmetric cranial trait (Hauser and De Stefano 1989) and there may be genetic factors involved (Saunders and Popovich 1978). Regional incidences of sella bridging range from 3.9% (Japanese male sample) to 34.9% in Canadian Iroquois. Little data are available on fossil hominins, except for brief comment (Washburn and Howell 1952), and there are few studies of more recent archeological material (Burrows et al. 1943). Further studies on this cranial region, and indeed on the area of the basicranial axis in general, could clearly be revealing.

6.9
The Bony Labyrinth

The temporal bone is no less neglected than the sphenoid in studies on earlier populations. The labyrinth, which is roughly of adult size at birth, is enclosed within the petrous bone of the temporal. In the primates, the size of the semicircular canals and the cochlea can be correlated with body mass (Spoor and Zonneveld 1999). Currently, the structures are best revealed by CT scans, when morphology is sufficiently clear to enable measurements to be taken. This enabled the bony labyrinth of Neanderthalers to be compared with other hominins, with the result that their closest affinities appeared to be with European Middle Pleistocene fossils, while Holocene groups appear to be more closely related to Asian and African *H. erectus* groups (Spoor et al. 2003).

Of a different level of enquiry is the detection of evidence of otosclerosis in ancient temporals. This disabling condition is today far more common in females, with progressive deafness usual. Genetic factors appear to be involved, and ethnic differences occur. A detailed study of a large sample of native American skulls revealed no evidence of otosclerosis, but further ethnic studies in relation to the past are needed (Gregg et al. 1965). In contrast to the otosclerosis evidence, mastoid sinus changes suggestive of infection were clearly in evidence.

6.10
Variation in the Postcranial Skeleton

During growth and living into old age, the skeleton reacts to a great variety of stresses. These may be superimposed over early established, even intrauterine, asymmetries. The nature of these asymmetries has by no means been well studied, and although of no great clinical relevance, it is of interest to skeletal biologists researching earlier populations. For instance, if a proportion of medieval battle victims are considered to have been well-trained archers, what skeletal features would help to indicate which of the bodies are the archers? Where military or sports training is carried out carefully, then bone and soft-tissue responses will be unnoticed. However, where there are training errors, growth anomalies, environmental problems or inadequate equipment, then overuse injuries can occur (Hutson 1990) and must have done in the past. The radiology of such injuries is well known today (Bowerman 1977).

Occupational or activity-related changes in the skeleton have been increasingly discussed over the past two decades (Capasso et al. 1999; Eckhardt 2000; Larsen 1997). Skeletal features as divergent as mastoid hypertrophy, costoclavicular sulcus, humeral hypertrophy and asymmetry, anconeus enthesopathy, and pilasterism, to name a few, have been discussed. Unfortunately, radiographic support for much of this research is limited, and there is a real need for specific radiological studies on the living and controlled extrapolation to peoples of the past. Radiological studies on industrial biomechanical stress provides sufficient evidence to argue in favor of more investigations of this kind. The increasing occurrence with time, of

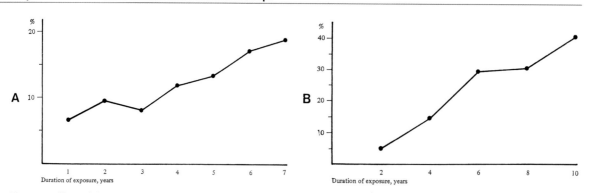

Fig. 6.12. Effect of duration of exposure in motor-saw operators, on the development of pseudocysts (**a**) and chronic atrophy (**b**) in the carpal area. After Horváth (1980)

pseudocysts and chronic atrophy in the carpal bones of motor-saw operators (Fig. 6.12), is the kind of x-ray evidence we need more of (Horváth 1980). A different approach to the study of both body form and activity was presented by Tanner (1964) in a consideration of Olympic athletes. Radiological comparisons of the athletes, in terms of muscle and bone dimensions at specific points in the body, revealed considerable variation in bone widths for the arms and legs (Fig. 6.13). Now that CT scanning is available, there is considerable potential for exploring the small-scale surface modifications, such as the enthesopathies, in far more detail.

6.11
Variation in Cortical Bone

The cortex of bones can vary in two major respects, namely, in thickness from the medullary cavity to the outer bone surface, or in terms of bone density as recorded for instance as dry weight or ash weight per cubic centimeter. Generally there is increase in both variables during childhood growth and commonly changes in later adult life, especially at a postmenopausal period in females. A range of diseases, from chronic anemia to muscular dystrophy, can affect cortical bone at various life stages.

Studies on cortical bone have produced a vast literature, as reviewed by Virtama and Helelä (1969) and others. Hormonal factors, nutrition, disease, and genetic (ethnic) factors are all influences to be taken into account. Physical activity can play a part in delaying cortical reduction. In the evaluation of density at a postmortem level, taphonomic factors must be taken into account (Bell et al. 1996), as these can significantly modify true density.

The degree of age variation in cortical thickness for normal individuals is shown in Fig. 6.14 (Virta-

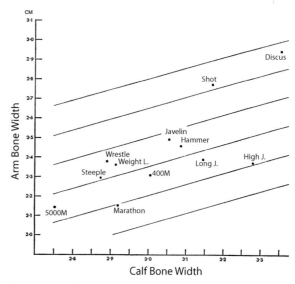

Fig. 6.13. Arm bone widths in relation to calf bone (tibia) widths in Olympic athletes. Modified from Tanner (1964)

ma and Helelä 1969). Variation in a medieval series of femora from Winchester, England (UK) is shown in Fig. 6.15 (Brothwell et al. 1968). It is now known that modeling changes occur during the adult period, with probably a similar pattern in both genders (Feik et al. 2000), so further studies on archeological series must take account of age in their analyses.

CT scanning has provided a further method of analyzing the structural and biomechanical variation seen in both fossil and more recent populations (Bridges et al. 2000). Interest in age-related cortical bone loss now spans four decades, and includes studies on prehistoric Amerindian series (Carlson et al. 1976; Perzigian 1973) and early historic Europeans (Mays 2006; Mays et al. 2006).

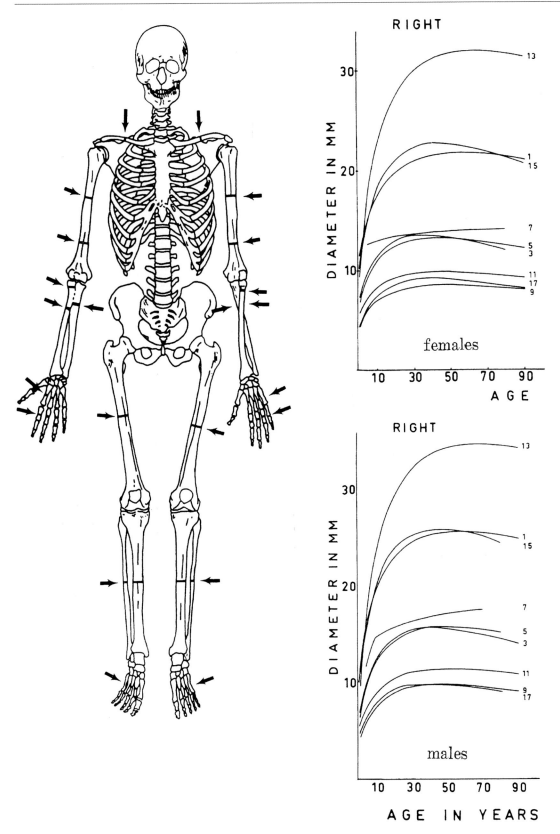

Fig. 6.14. Cortical thickness variation in a normal European population. **a** Sites of measurements taken by Virtama and Helelä (1969). **b** Transverse diameters (*right*) for various skeletal positions for males and females

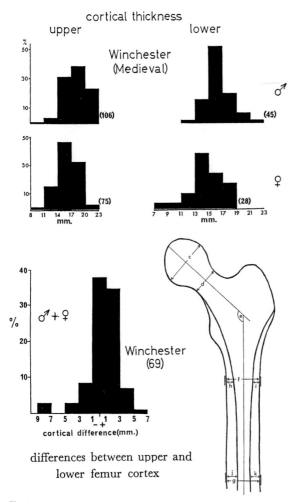

cortical thickness

differences between upper and lower femur cortex

Fig. 6.15. Variation in cortical thickness of the upper femur shaft in a medieval population from Winchester, England (UK). All adults

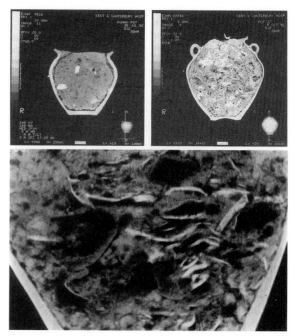

Fig. 6.16. CT scan view of bone within cremation vessels, as studied by Anderson and Fell (1995)

of bone preserved, so that in some instances it may not be worth spending further time on its analysis (Anderson and Fell 1995).

6.12
Cremations

Finally, mention should be made as regards the value of x-rays in the study of cremations. While measurements of cortical bone are of no value because of bone shrinkage and deformity, there is still value in exploring the pathology present in some cremated material. In particular, healed fractures, Harris lines, and even apical infection of the jaws could be revealed by radiographic techniques. In the case of cremations packed into funerary containers, CT scans enable the skeletal remains to be analyzed in situ (Fig. 6.16). The advantage of this is that bone in many cremations is fissured and fragile, and its removal from cremation urns results in bone disintegration. A CT scan may also indicate the extent

References

Acsádi G, Nemeskéri J (1970) History of Human Life Span and Mortality. Akadémiao Kiadó, Budapest

Anderson T, Fell C (1995) Analysis of Roman cremation vessels by computerized tomography. J Archaeol Sci 22:609–617

Arambourg C (1963) Le gisement de Ternifine, II L'Atlanthropus mauritanicus. Arch L'Inst Paleo Humaine, Mem 32. Masson, Paris, pp 37–190

Bang G (1989) Age changes in teeth: developmental and regressive. In: Iscan MY (ed) Age Markers in the Human Skeleton. Thomas, Springfield, pp 211–235

Bell LS, Skinner MF, Jones SJ (1996) The speed of post mortem change to the human skeleton and its taphonomic significance. Forensic Sci Int 82:120–140

Black D (1929) Preliminary note on additional Sinanthropus material discovered in Choukoutien during 1928. Bull Geol Soc China 8:15–20

Bowerman JW (1977) Radiology and Injury in Sport. Appleton-Century-Crowfts, New York

Bridges PS, Blitz JH, Solano MC (2000) Changes in long bone diaphyseal strength with horticultural intensification in west-central Illinois. Am J Phys Anthropol 112:217–238

Brothwell DR, Molleson T, Metreweli C (1968) Radiological aspects of normal variation in earlier skeletons: an exploratory study. In: Brothwell DR (ed) The Skeletal Biology of Earlier Human Populations. Pergamon, Oxford, pp 149–172

Brown B, Walker A (1993) The dentition. In: Walker A, Leakey R (eds) The Nariokotome *Homo erectus* skeleton. Harvard University Press, Cambridge, pp 161–192

Brown T (1973) Morphology of the Australian skull. Australian Aboriginal Studies No 49, Canberra

Bruwelheide KS, Beck J, Pelot S (2001) Standardized protocol for radiographic and photographic documentation of human skeletons. In: Williams E (ed) Human Remains: Conservation, Retrieval and Analysis. BAR (International), Oxford, Ser 934, pp 153–165

Buckland-Wright JC (1970) A radiographic examination of frontal sinuses in early British populations. Man 5:512–517

Burrows H, Cave AJE, Parbury K (1943) A radiographical comparison of the pituitary fossa in male and female whites and negroes. Br J Radiol 16:87

Buxton LHD (1928) Human remains. In: Garrod DAE, Buxton LHD, Smith GE, Bate DMA (eds) Excavation of a Mousterian rock shelter at Devils Tower, Gibraltar. J Roy Anthrop Inst 58:57–85

Capasso L, Kennedy KAR, Wilczak CA (1999) Atlas of Occupational Markers on Human Remains. Edigrafital SpA, Teramo

Carlson DS, Armelagos GJ, Van Gerven DP (1976) Patterns of age-related cortical bone loss (osteoporosis) within the femoral diaphysis. Hum Biol 48:295–314

Conroy GC, Vannier MW (1985) Endocranial volume determination of matrix-filled fossil skulls using high-resolution computed tomography. In: Tobias PV (ed) Hominid Evolution: Past, Present and Future. Liss, New York, pp 419–426

Cook DC, Buikstra JE, DeRousseau CJ, Johanson DC (1983) Vertebral pathology in the Afar Australopithecines. Am J Phys Anthropol 60:83–101

Day MH, Molleson TI (1973) The Trinil femora. In: Day MD (ed) Human Evolution. Taylor and Francis, London, pp 127–154

Eckhardt R B (2000) Human Paleobiology. University Press, Cambridge

El-Nofely AA, Iscan MY (1989) Assessment of age from the dentition in children. In: Iscan MY (ed) Age markers in the human skeleton. Thomas, Springfield, pp 237–254

Fairgrieve SI, Bashford J (1988) A radiographic technique of interest to physical anthropologists. Am J Phys Anthropol 77:23–26

Feik SA, Bruns TW, Clement JG (2000) Regional variations in cortical modelling in the femoral mid-shaft: sex and age differences. Am J Phys Anthropol 112:191–205

Gorjanović-Kramberger D (1906) Der diluviale Mensch von Krapina in Kroatien. Ein Beigrag zur Paläoanthropologie. Kreidel, Wiesbaden

Gregg JB, Holzhueter AM, Steele JP, Clifford S (1965) Some new evidence on the pathogenesis of otosclerosis. Laryngoscope 75:1268–1292

Hauser G, De Stefano GS (1989) Epigenetic Variants of the Human Skull. Schweizerbart, Stuttgart

Horváth F (1980) X-ray morphology of occupational locomotor diseases. University Park Press, Baltimore

Hutson MA (1990) Overuse injuries of the ankle and foot. In: Hutson MA (ed) Sports Injuries, Recognition and Management. University Press, Oxford, pp 152–162

Jacob T (1967) The racial history of the Indonesian region. Neerlandia, Utrecht

Kallay J (1963) A radiographic study of the Neanderthal teeth from Krapina, Croatia. In: Brothwell DR (ed) Dental Anthropology. Pergamon, London, pp 75–86

Kisling E (1966) Cranial Morphology in Down's Syndrome. Munksgaard, Copenhagen

von Koenigswald GHR (ed) (1958) Neanderthal Centenary 1856–1956. Zoon, Utrecht

Koppe T, Nagai H (1999) Factors in the development of the paranasal sinuses. In: Koppe T, Nagai H, Alt KW (eds) The paranasal sinuses of higher primates. Quintessence, Chicago, pp 133–149

Krogman WM (1962) The Human Skeleton in Forensic Medicine. Thomas, Springfield

Larsen CS (1977) Bioarchaeology, Interpreting Behaviour from the Human Skeleton. University Press, Cambridge

Leakey RE, Leakey MG, Behrensmeyer AK (1978) The hominid catalogue. In: Leakey MG, Leakey RE (eds) Koobi Fora Research Project. Clarendon, Oxford, pp 86–187

Le Gros Clark WE (1964) General features of the Swanscombe skull bones. In: Ovey CD (ed) The Swanscombe Skull. A Survey of Research on a Pleistocene Site. Royal Anthropological Institute, London, pp 135–137

Lumley M-A de (1981) Les anténéandertaliens en Europe. In: Sigmon BA, Cybulski JS (eds) *Homo erectus*, Papers in Honor of Davidson Black. University of Toronto Press, Toronto, pp 115–132

Lumley-Woodyear M-A de (1973) Anténéandertaliens et néandertaliens du bassin méditerranéen occidental Européen. Études Quaternaires 3, Marseille

Mays SA (2006) Age-related cortical bone loss in women from 3rd-4th century AD population from England. Am J Phys Anthropol 129:518–528

Mays S, Turner-Walker G, Syversen U (2006) Osteoporosis in a population from medieval Norway. Am J Phys Anthropol 131:343–351

Moss ML (1971) Ontogenetic aspects of cranio-facial growth. In: Moyers RE, Krogman WM (eds) Cranio-Facial Growth in Man. Pergamon, Oxford, pp 109–124

Neubauer S, Gunz P, Mitteroecker P, Weber GW (2004) Three-dimensional digital imaging of the partial Australopithecus africanus endocranium MLD 37-38. Can Assoc Radiol J 55:271–278

Pawson IG, Huicho L, Muro M, Pacheco A (2001) Growth of children in two economically diverse Peruvian high-altitude communities. Am J Hum Biol 13:323–340

Perzigian AJ (1973) Osteoporotic bone loss in two prehistoric Indian populations. Am J Phys Anthropol 39:87–95

Price JL, Molleson TI (1974) A radiographic examination of the left temporal bone of Kabwe man, Broken Hill mine, Zambia. J Arch Sci 1:285–289

Radovčić J, Smith FH, Trinkaus E, Wolpoff MH (1988) The Krapina hominids. Croatian Natural History Museum, Zagreb

Saunders E (1837) The teeth a test of age, considered with reference to the factory children. Renshaw, London

Saunders SR, Popovich F (1978) A family study of two skeletal variants: atlas bridging and clinoid bridging. Am J Roent 49:193–203

Schuster FP, Finnegan M (1977) Racial distance: a multivariate analysis of roentgengraphic measurements in Eskimos, Indians and Whites. HOMO 28:227–235

Schüller A (1943) Note on the identification of skulls by X-ray pictures of the frontal sinuses. Med J Australia 1:554–556

Seidler H, Falk D, Stringer C, Wilfing H, Müller GB, Nedden D, Weber GW, Reicheis W, Arsuaga J-L (1997) A comparative study of stereolithographically modelled skulls of Petralona and Broken Hill: implications for future studies of middle Pleistocene hominid evolution. J Hum Evol 33:691–703

Singer R (1958) The Rhodesian, Florisbad and Saldanha skulls. In: von Koenigswald GHR (ed) Neanderthal Centenary 1856–1956. Zoom, Utrecht, pp 52–62

Skinner MF, Sperber GH (1982) Atlas of radiographs of early man. Liss, New York

Sorg MH, Andrews RP, Iscan MY (1989) Radiographic aging of the adult. In: Iscan MY (ed) Age Markers in the Human Skeleton. Thomas, Springfield, pp 169–173

Spoor F, Hublin J-J, Braun M, Zonneveld F (2003) The bony labyrinth of Neanderthals. J Hum Evol 44:141–165

Spoor F, Stringer C, Zonneveld F (1998) Rare temporal bone pathology of the Singa calvaria from Sudan. Am J Phys Anthropol 107:41–50

Spoor F, Zonneveld F (1999) Computed tomography-based three-dimensional imaging of hominid fossils: features of the Broken Hill 1, Wadjak 1, and SK 47 crania. In: Koppe T, Nagai H, Alt KW (eds) The Paranasal Sinuses of Higher Primates. Quintessence Publishing, Chicago, pp 207–226

Szilvássy J (1982) Zur variation, Entwicklung und Vererbung der Stirnhöhlen. Ann Naturhist Mus Wien 84:97–125

Tanner JM (1964) The Physique of the Olympic Athlete. Allen and Unwin, London

Thompson JL, Illerhaus B (2000) CT reconstruction and analysis of the Le Moustier 1 Neanderthal. In: Stringer CB, Barton RNE, Finlayson JC (eds) Neanderthals on the Edge. Oxbow, Oxford, pp 249–255

Thorne AG (1971) Mungo and Kow Swamp: morphological variation in Pleistocene Australians. Mankind 8:85–91

Tobias PV (1968) The pattern of venous sinus grooves in the robust Australopithecines and other fossil and modern Hominids. In: Anthropologie und Humangenetik. Fischer, Stuttgart, pp 1–10

Trinkaus E (1993) Femoral neck-shaft angles of the Qafzeh-Skhul early modern humans, and activity levels among immature Near Eastern middle paleolithic hominids. J Hum Evol 25:393–416

Ubelaker DH (1989) The estimation of age at death from immature human bone. In: Iscan M Y (ed) Age Markers in the Human Skeleton. Thomas, Springfield, pp 55–70

Virtama P, Helelä T (1969) Radiographic measurements of cortical bone. Acta Radiol Supplement 293

Vlček E (1967) Die sinus frontales bei europäischen Neandertalern. Anthrop Anz 30:166–189

Walker RA, Lovejoy CO (1985) Radiographic changes in the clavicle and proximal femur and their use in determination of skeletal age at death. Am J Phys Anthropol 68:67–78

Washburn SL, Howell FC (1952) On the identification of the hypophyseal fossa of Solo man. Am J Phys Anthropol 10:13

Weidenreich F (1935) The Sinanthropus population of Choukoutien (locality 1) with a preliminary report on new discoveries. Bull Geol Soc China 14:427–461

Whittaker D (2000) Ageing from the dentition. In: Cox M, May S (eds) Human Osteology in Archaeology and Forensic science. Greenwich Medical Media, London, pp 83–99

Wind J, Zonneveld F (1985) Radiology of fossil hominid skulls. In: Tobias PV (ed) Hominid Evolution: Past, Present and Future. Liss, New York, pp 437–442

Woo J-K, Peng R (1959) Fossil human skull of early paleoanthropic stage found at Mapa, Shaoquan, Kwangtung province. Vert Palas 3:176–183

Zonneveld FW, Wind J (1985) High resolution computed tomography of fossil hominid skulls: a new method and some results. In: Tobias PV (ed) Hominid Evolution: Past, Present and Future. Liss, New York, pp 427–436

Concluding Comments

Rethy K. Chhem and Don R. Brothwell

In this volume, we have attempted to review as broadly as possible the application of radiographic techniques to the study of the organic remains associated with past cultures. At present, there has been far more applications to the study of human remains, and far less to animal or plant material. However, with a growing knowledge of the value of radiographic study of archaeological objects, both organic and inorganic, it could well be that the emphasis might change significantly. In particular, we would predict that there would be far more studies on animal bones and teeth, in terms of both normal variation and veterinary paleopathology. The study of human fossils has significantly neglected radiographic techniques, and we would hope that in the future, all new fossil discoveries and descriptions would include radiological data. The archaeological remains of organic material present special problems in terms of the far greater variation in the degree of preservation. Plant remains can be carbonized and fused into masses. Animal bones and teeth can be cooked, butchered, and thrown into refuse pits. Varying states of fossilization may radically change the mineral content, and acid bogs may significantly demineralize bones and teeth. Cremated bones may be in jars together with burial matrix.

New technology or improved machines may offer additional ways of investigating the past. Xeroradiography is a part of the past, but micro-computed-tomography (micro-CT) scanning opens up a much smaller world for investigation. Could the latter be of special value in evaluating what part of a coprolite should be rehydrated for microscopic analysis? Insect damage to cereal grain, or even the insect remains themselves might yield further taxonomic information with the aid of micro-CT.

As x-rays can be damaging at the cellular level (especially for DNA), there is a need to avoid too much radiation of specimens. There is thus a need to share radiographic information with colleagues in order to reduce repeated x-raying of the same specimen. Indeed, there is a good argument for the archiving of digital radiographic images of ancient material either at university centers or museum archival centers.

Costs will remain a problem to many without the direct support of a radiology department. Where budget is limited, as in nonmedical universities and museum departments, the purchase and use of a small digital x-ray unit will be ideal for small- to medium-sized objects, but mummies and mammoth bones demand something larger. It need hardly be said that it is important to have associated with the x-ray, relevant details of the site, period, culture, and reason for the x-ray. Ideally, publications should also record where the radiographic records are curated. If human remains are reburied, then it is especially important to curate for the future any radiographic data. Old radiographs should also be inspected for deterioration, and when possible digitized and stored on disc.

While at a clinical level methodologies are in place to obtain ideal x-rays of patients in relation to specific health problems, there is still a need with many archaeological specimens to standardize positions and orientations; for instance when studying cortex thickness at specific sites in ancient skeletons or evaluating cranial thickness at specific positions of the skull. There are as yet remarkably few radiographic studies specifically on archaeological material. This is not because there is no value in them, but because there is as yet no "mind set" in archaeology, which stimulates work of this kind.

This book has also tried to address a major weakness in the method of paleopathology in which the differential diagnosis of skeletal diseases is rarely discussed. For that purpose, we have dedicated a special chapter on diagnostic paleoradiology to help nonradiologists to approach the diagnosis of lesions in dry bones from the archaeological record. However, that chapter did not aim at teaching paleopathologists to read radiographic films, but instead to expose them to the method used by skeletal radiologists in their reasoning process during the interpretation of x-ray findings.

We hope this volume will help to change attitudes toward the radiological approach to bioarchaeological materials. There are certainly several journals, such as the Journal of Archaeological Science and the

International Journal of Osteoarchaeology, which would consider such work and engage radiologists in the review and evaluation of manuscripts containing x-ray studies. Also, where a substantial study has been made that has relevance to both the archaeological and radiological communities, articles may be needed in both types of journals. Even publications in clinical radiology, agricultural history, palaeobotany, zooarchaeology, and veterinary journals might at times be relevant. As science expands into studies of our past, it is essential that radiological studies are not left behind.

Subject Index